Happy hunt.

&

INSECTS OF .LAND

A FIELD GUIDE

DR STEPHEN McCORMACK, a professional entomologist, has published widely on Irish beetles, wasps, butterflies, bugs, spiders and snails. One of Ireland's top insect experts, he teaches on field courses and in schools.

DR EUGENIE REGAN, a professional ecologist at the UNEP World Conservation Monitoring Centre in Cambridge and formerly at the National Biodiversity Data Centre, has published widely on Ireland's wildlife. A regular contributor to radio shows, she has previously published an introductory guide to Irish butterflies and other insects.

CHRIS SHIELDS, one of the world's leading natural history illustrators and award-winning artist, has produced over 20,000 wildlife illustrations in over 300 books. His work is published by A&C Black, BBC *Wildlife* Magazine, HarperCollins, Random House and some environmental charities such as WWF and the RSPB.

The Collins Press

This book is dedicated to Irish entomologists: past, present, and future.

First published in 2014 by
The Collins Press
West Link Park
Doughcloyne
Wilton
Cork

Reprinted 2016

A CIP record for this book is available from the British Library.

ISBN: 978-1-84889-208-8
PDF ISBN: 978-1-84889-521-8
ePub ISBN: 978-1-84889-522-5
Kindle ISBN: 978-1-84889-523-2

Design and typesetting by Paul Whelan (Lichens.ie)
Typeset in Adobe Garamond Pro
Printed in Poland by Białostockie Zakłady Graficzne SA

Contents

Alphabetical species list

Bumblebees

Grasshoppers and allies

Earwigs

Shield Bugs

Acknowledgements

We have been inspired, encouraged, and supported in our passion for entomology by many naturalist friends and colleagues including Roy Anderson, Ken Bond, Tom Brereton, Úna Fitzpatrick, Garth Foster, Richard Fox, Jervis Good, Leo Hallisey, Jesmond Harding, Derek Lott, Liam Lysaght, Kieran McCarthy, Ian Middlebrook, Brian Nelson, Myles Nolan, Áine O'Connor, Jim O'Connor, Chris Peppiatt, Cilian Roden, Constanti Stefanescu, Peter Sutton, Chris van Swaay, Chris Wilson, and Stephen Ward. Thank you for your support and encouragement.

We would like to sincerely thank Paul Whelan for his enthusiasm, guidance, and design of this book.

Thanks also to Lily Kate and Oisín McCormack for their patience and inspiration in writing this book.

Stephen McCormack and Eugenie Regan
2014

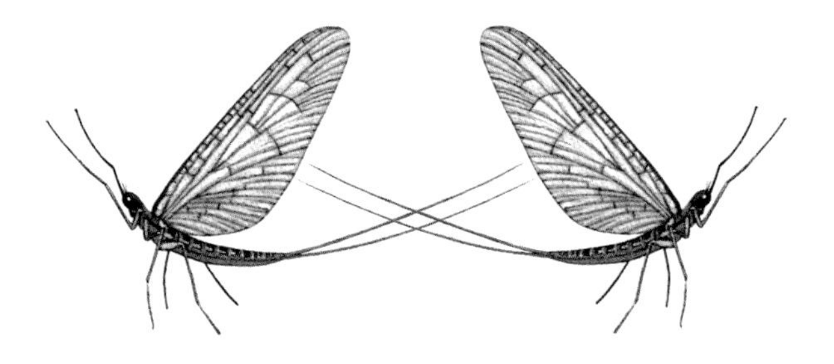

Introduction

This book is an introduction to the fascinating, hidden world of Irish insects. Insects are the most diverse group of animals in Ireland, yet they are an overlooked and neglected part of our fauna. No other group of living creatures has such variety of form, colour, function, and habitat. Insects make up over one third of the total known number of all Irish species and are the most species-rich group of organisms on earth, constituting over 60 per cent of the roughly 1.5 million described species. The total number of known insect species for Ireland is over 11,400 and many more have yet to be discovered.

The basic body plan of insects lends itself to virtually endless modification and adaptation and insects occupy almost all terrestrial and freshwater habitats on earth. This great diversity creates a barrier to studying insects – there are just too many different types to study without specialist knowledge and many are just too small to see properly without magnifying equipment.

This book provides an introduction to seven Irish insect groups – butterflies, dragonflies, ladybirds, bumblebees, shield bugs, and grasshoppers and earwigs – a total of 128 species that are all almost easily identified without magnification. All Irish species of these groups are illustrated and described with the exception of rare migratory species. A vice-county distribution map is given for each species which presents up-to-date information on the known distribution of each species. For ladybirds, shield bugs and grasshoppers the known distribution may be different from their actual distributions, reflecting the lack of systematic recording of these groups to date. Hopefully, the information presented in this book will facilitate improved recording of the distributions of these insects in Ireland.

There is also a general overview of Irish insects and we hope that it fuels your appetite to delve further into the natural history of this animal group.

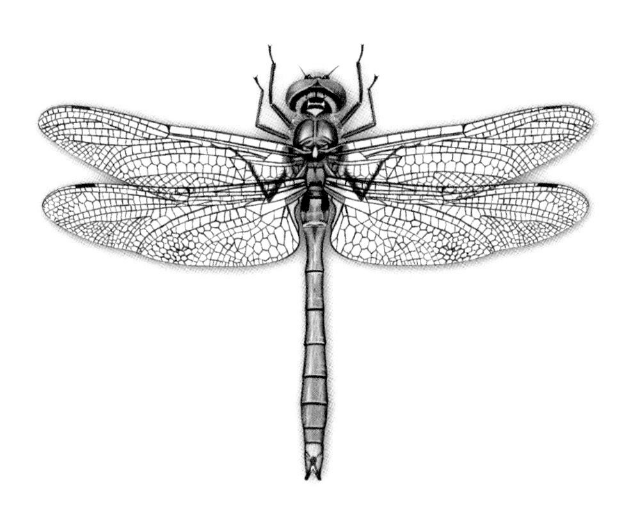

Ireland's insects

Insects come in many shapes and forms, from beetles to bugs and from earwigs to butterflies. A standard definition is that an insect has six legs and the adult body is divided into three distinct parts: the head, thorax, and abdomen. Insects usually have two pairs of wings but there are many instances where wings are modified or absent. For example, in adult beetles the first pair of wings is modified into a hardened wing case, serving as protection for the second pair which may be used for flight. Scientists divide the insects into related groups or 'orders' of which there are 21 represented in Ireland. These are briefly introduced below.

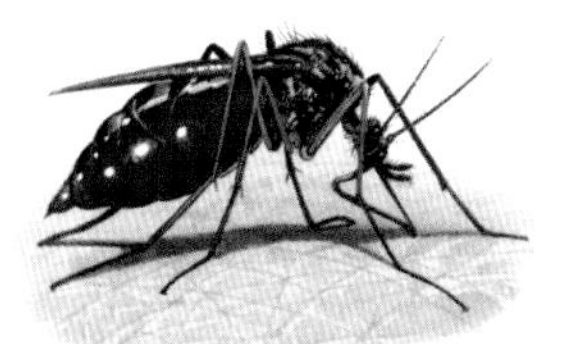

A female *Anopheles* mosquito feeding on blood. The aquatic larvae live in tree holes.

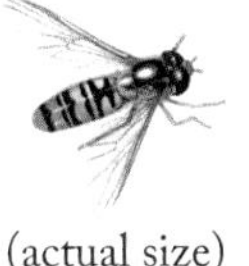

Marmalade Fly (*Episyrphus balteatus*), is a very common hoverfly whose larvae are predators of aphids.

(actual size)

1. Flies (Order Diptera)

This is the largest insect group in Ireland with over 3,300 known Irish species. They are minute to large insects in which the second pair of wings is reduced to small, knobbed appendages, leaving only one pair of wings. They have sucking mouthparts that are frequently adapted for piercing, as is the case in the mosquitoes. Many flies mimic the black and yellow of bees and wasps, but resemblances are superficial and closer examination will reveal only two wings, clearly indicating a true fly. There are an estimated 240,000 species of fly in the world.

The Common Wasp (*Vespa vulgaris*)

A worker of the familiar Honey Bee (*Apis mellifera*)

A worker of the Black Ant (*Lasius niger*)

2. Bees, wasps, ants, parasitoid wasps & sawflies (Order Hymenoptera)

This diverse insect group comes a close second to the flies in terms of species numbers with over 3,200 known Irish species. Bees constitute just over 100 species, ants over 20 species, while the parasitoid wasps and sawflies are by far the most species-rich with over 3,000 species. The Hymenoptera are minute to large insects, usually with two pairs of membranous wings of which the front pair is the larger. Over 150,000 species are recognised globally, with many more remaining to be described.

(actual size)

The soldier beetle (*Rhagonycha fulva*) is abundant everywhere on flowers in summer.

3. Beetles (Order Coleoptera)

Beetles are the third most species-rich group in Ireland with over 2,100 known species. They are minute to large insects, normally with two pairs of wings of which the front pair are hardened. The hindwings are membranous and usually folded away out of sight beneath the forewings. About 40 per cent of all described insect species are beetles (approximately 400,000 species).

Ground Beetle (*Abax parallellepipedus*)

(actual size)

Buff Arches moth (*Habrosyne pyritoides.* Caterpillars of the Buff Arches feed only on Bramble.

4. Butterflies & moths (Order Lepidoptera)

In Ireland, there are over 1,400 known species, the vast majority of which are moths. The Lepidoptera are minute to large insects, usually with two pairs of membranous wings that are more or less covered with tiny scales. It is estimated that there are over 170,000 species worldwide. Caterpillars mostly feed on plant material and each species consumes only certain plant species.

6-spot Burnet Moth (*Zygaena filipendulae*) is found in flowery meadows with bird's-foot-trefoil.

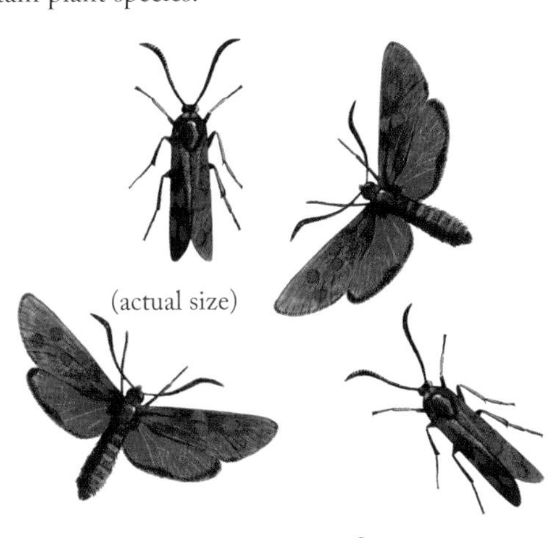

(actual size)

5. True bugs (Order Hemiptera)

The true bugs are minute to large insects of widely differing shapes and habits but all possessing piercing mouthparts adapted for sucking the juices of plants and other animals. There are over 770 known species in Ireland, including shield bugs and aphids, and over 50,000 species worldwide.

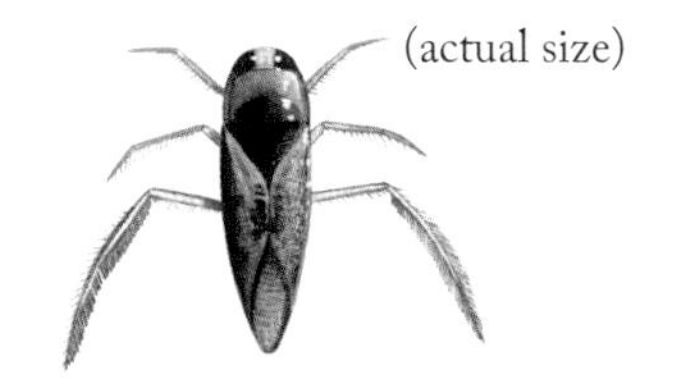

The Backswimmer or Water Boatman (*Notonecta glauca*) is a predatory aquatic bug.

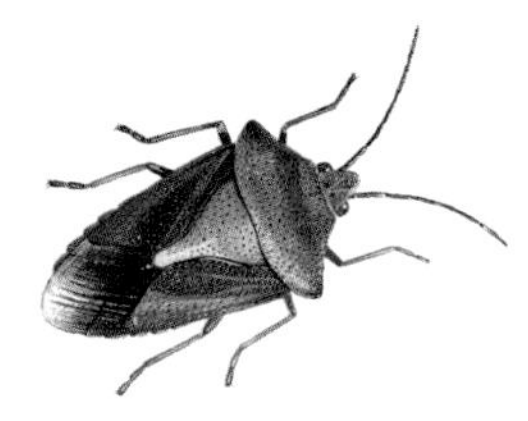

The Hawthorn Shield Bug (*Acanthosoma haemorrhoidale*) is a herbivorous bug.

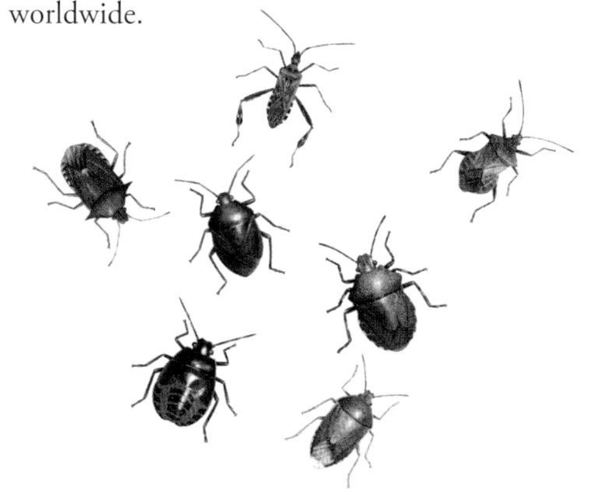

6. Caddis flies (Order Trichoptera)

Small, medium and large insects with two pairs of wings covered in tiny hairs. Caddis flies are structurally very similar to certain moths and some members of the two groups are easily confused. However, the hairy (not scaly) wings and the lack of a coiled proboscis will distinguish the caddis flies. Caddis fly larvae are aquatic and many species make protective cases of sand, gravel, twigs or other debris. There are 147 species recorded from Ireland and approximately 12,000 described species worldwide.

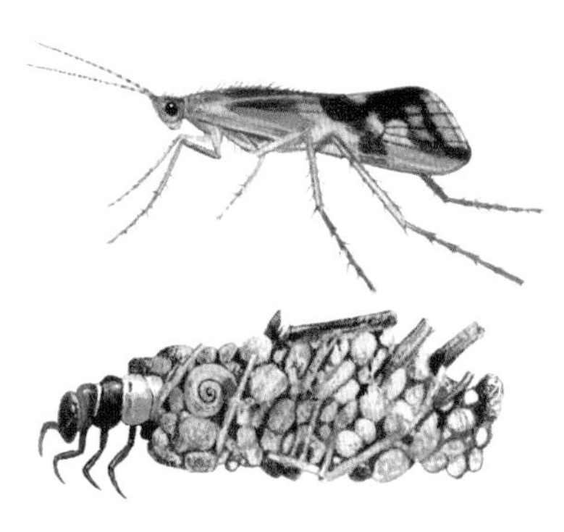

Larvae of the Cinnamon Sedge (*Limnephilus lunatus*) live in rivers, lakes and streams. The adults are a favourite food of trout.

7. Lice (Order Phthiraptera)

The Head Louse (*Pediculus humanus*) is solely parasitic on humans and the producer of the nit.

(actual size)

All lice are external parasites of mammals and birds. There are 117 known Irish species and over 3,000 species worldwide. The name Phthiraptera is derived from the Greek 'phthir' meaning lice and 'aptera' meaning wingless. The literal translation, wingless lice, is appropriate for all members of the order.

8. Booklice & barklice (Order Psocoptera)

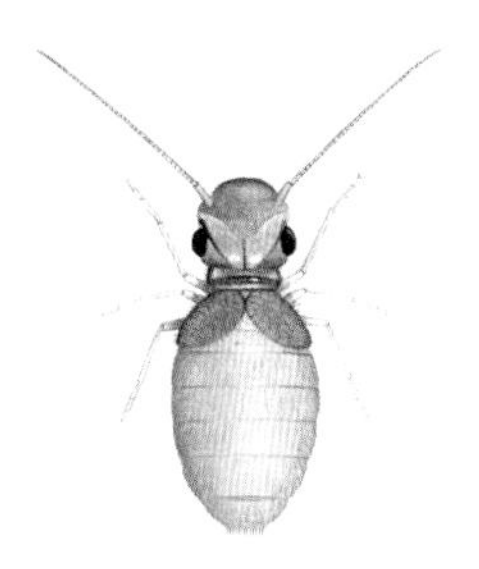

Booklice like this *Lepinotus* are frequent in damp places in houses.

Barklice generally live in moist terrestrial environments (in leaf litter, beneath stones or under the bark of trees) where they forage on algae, lichens, fungi and various plant products. They may grow to 10mm in length. Booklice are more common in human dwellings and warehouses, are wingless and much smaller than barklice (less than 2mm). Most species feed on stored grains, bookbindings, wallpaper paste, fabric sizing, and other starchy products. There are 46 species recorded from Ireland and over 5,500 species described worldwide.

9. Fleas (Order Siphonaptera)

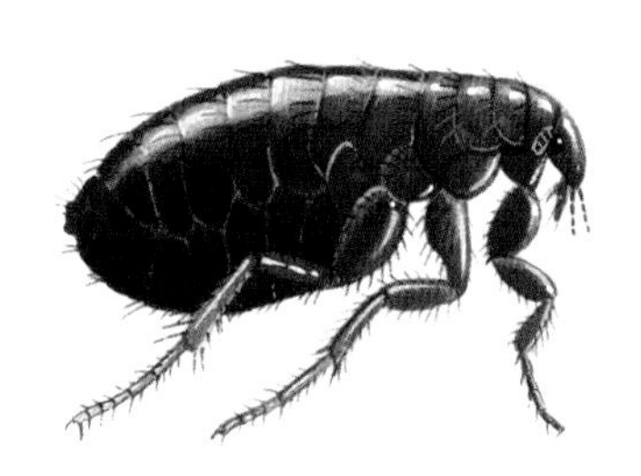

Cat Flea (*Ctenocerphalides felis*)

(actual size)

The name Siphonaptera is derived from the Greek words '*siphon*' meaning a tube or pipe and '*aptera*' meaning wingless. This is appropriate for these wingless insects whose mouthparts are adapted for piercing skin and sucking blood. All fleas are bloodsucking external parasites. There are 40 different flea species recorded from Ireland and over 2,000 species described.

Aeolothrips intermedius
(not recorded in Ireland)

10. Thrips (Order Thysanoptera)

The name Thysanoptera, derived from the Greek '*thysanos*' meaning fringe and '*ptera*' meaning wings, refers to the slender wings that bear a dense fringe of long hairs. Thrips are generally small insects (under 3mm). Most species feed on plant tissues (often in flower heads), but some are predators of mites and various small insects (including other thrips). There are 40 species known in Ireland and over 5,000 species described.

Common Darter *Sympetrum striolatum*

11. Dragonflies & damselflies (Order Odonata)

The Odonata are long, slender-bodied, carnivorous insects with very large compound eyes and two pairs of wings. There are 24 resident Irish species and some 5,900 species described worldwide. The juvenile stages develop in fresh water and are predators of other aquatic invertebrates. It is unlikely that dragonflies will be confused with any other insects.

The Pond Olive (*Cloeon dipterum*) is common near still water all over Ireland.

Mayfly nymph

12. Mayflies (Order Ephemeroptera)

The mayflies are a group of freshwater insects well known to fishermen. There are 33 known species in Ireland. They are small, medium and large-sized insects with two or three long 'tails' and one or two pairs of delicate wings. The hindwings, when present, are always considerably smaller than the front pair. These insects are usually found in the vicinity of water. There are about 2,500 species known worldwide.

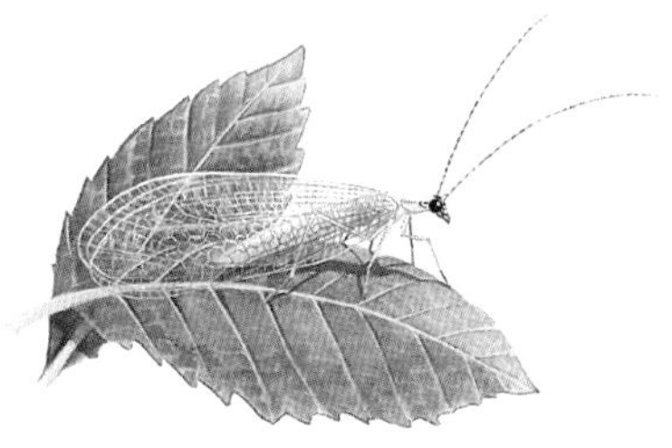

The lacewing (*Chysopa abreviata*) is a predatory insect that feeds on aphids.

13. Lacewings (Order Neuroptera)

The lacewings are soft-bodied insects that are generally brown or green. They have two pairs of flimsy wings covered with a delicate network of veins and held roof-wise over the body when at rest. Thirty-two species of lacewings have been recorded from Ireland and about 6,000 species known worldwide.

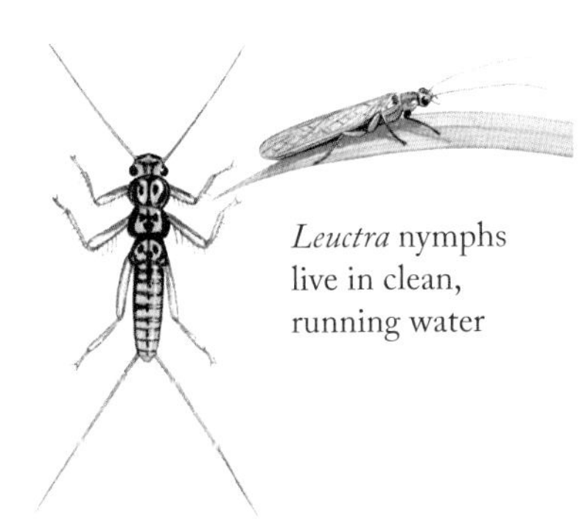

Leuctra nymphs live in clean, running water

14. Stoneflies (Order Plecoptera)

The stoneflies are aquatic insects with over 3,500 species worldwide and just 20 species in Ireland. They are medium-sized insects with two pairs of membranous wings and long, slender antennae. All species are intolerant of water pollution and their presence in a water body is usually an indicator of good or excellent water quality.

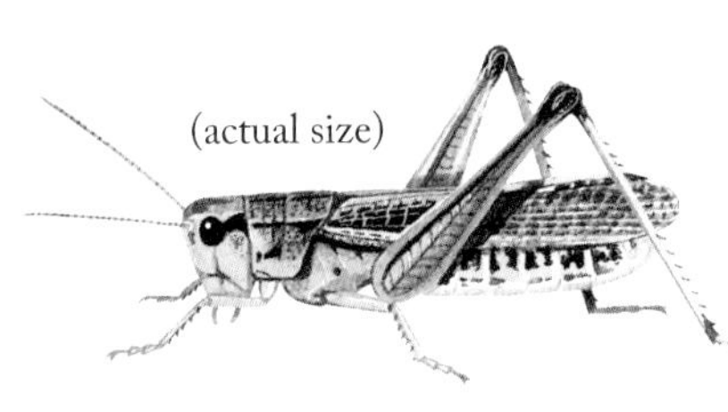

The Large Marsh Grasshopper (*Stethophyma grossum*) lives in wet bogs.

15. Grasshoppers & Crickets (Order Orthoptera)

These are easily recognised insects, with a shape so characteristic that it is impossible to confuse them with any other insects. They are medium to large insects with a stout body, large blunt head and enlarged hind legs that are modified for jumping. There are 12 Irish species and over 25,000 known species worldwide.

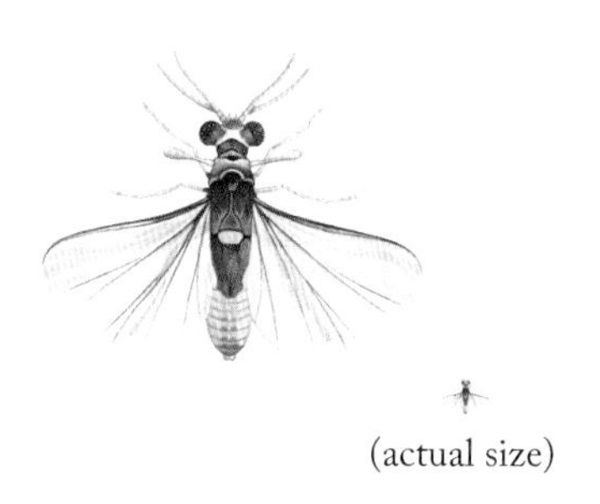

Elenchus tenuicornis

16. Stylopids (Order Strepsiptera)

The Strepsiptera are parasites of other insects, such as bees, wasps and silverfish. There are four Irish species and about 600 species known worldwide. These insects are very difficult to locate and one often has to find the host in order to find the female. They exhibit extreme sexual dimorphism: the adult males are free-living while the females are internal parasites of bees and bugs.

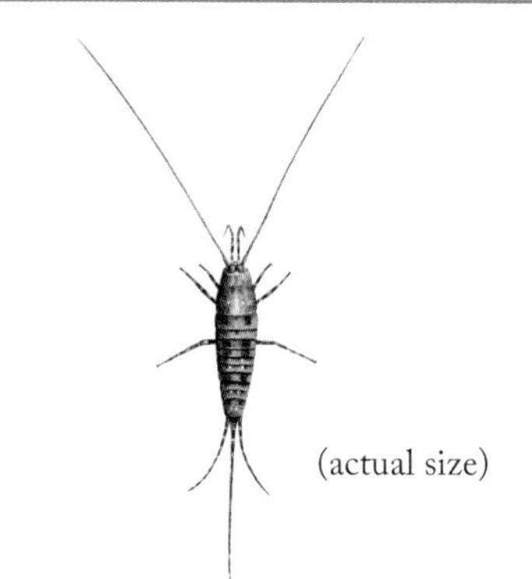

Petrobius maritimus is common near the coast all around Ireland.

17. Jumping bristletails (Archaeognatha)

The Archaeognatha are an order of wingless insects that are among the least evolutionarily changed. They are small with elongated bodies bent into an arch shape and three long tail-like structures. Four species of these insects are known from Ireland and there are only 350 species known worldwide.

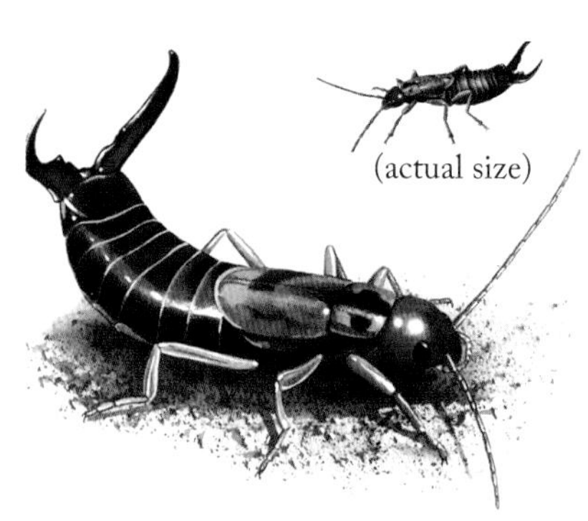

Male Common Earwig (*Forficula auriculata*) raising forceps in defence.

18. Earwigs (Order Dermaptera)

The earwigs are distinctive, elongate, brownish insects, usually with short, leathery forewings reaching only a short way down the body. The cerci (a pair of appendages at the end of the abdomen) are modified into stout forceps, strongly curved in the male. The general shape of these insects is such that they may be confused with certain of the beetles, but the forceps always distinguish the earwigs. There are three earwig species recorded from Ireland and about 2,000 species known worldwide.

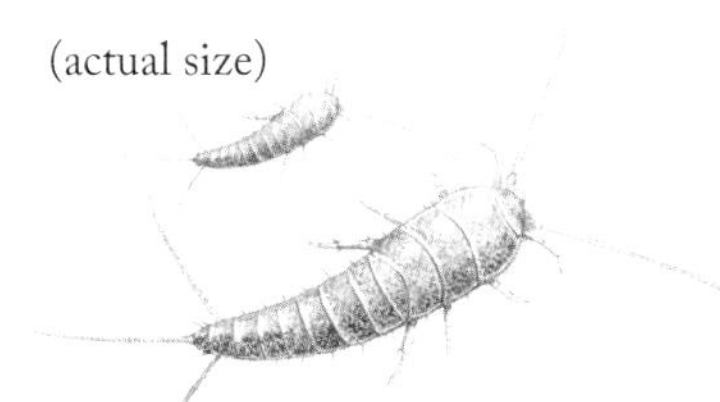

The Silverfish (*Lepisma saccharina*) is a common insect in houses

19. Bristletails/silverfish (Order Thysanura)

Just two species of these wingless insects are known from Ireland. Bristletails are up to 20mm long, with carrot-shaped bodies clothed with shiny scales and long, thread-like antennae with three long, segmented tail filaments at the rear. There are over 300 species known worldwide.

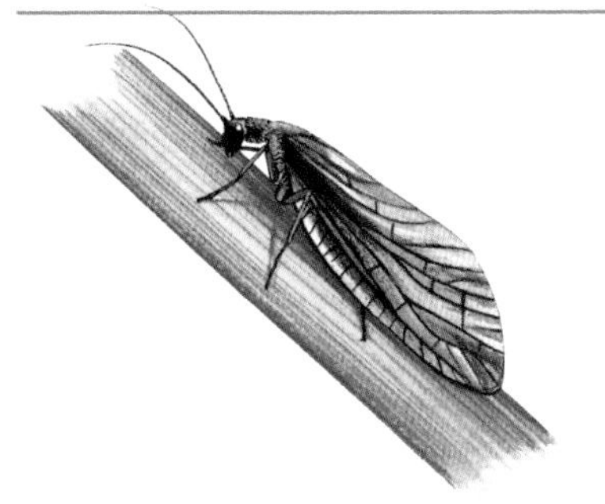

The aquatic larva of the Alderfly (*Sialis lutaria*) is a voracious predator on other insect larvae.

20. Alderflies (Order Megaloptera)

The alderflies are another small insect order with only two species recorded from Ireland. These insects are small, medium and large soft-bodied insects, generally brown or green. They have two similar pairs of flimsy wings, covered with a delicate network of veins, which are held roof-wise over the body at rest. There are about 300 species known worldwide.

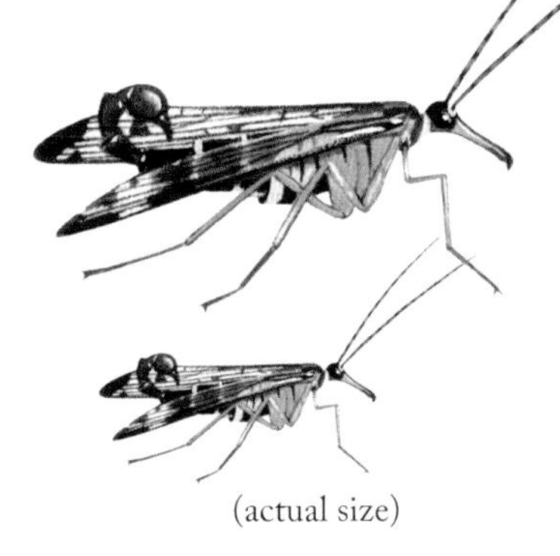

Only recorded twice in Ireland, the scorpion fly *Panorpa germanica* has not been seen in over 100 years.

21. Scorpion flies (Order Mecoptera)

Only one species of scorpion fly is known from Ireland and approximately 550 species worldwide. The scorpion flies are small- to medium-sized insects with slender, elongated bodies. They are called scorpion flies after their largest family, Panorpidae, in which the males have enlarged genitals that look similar to the stinger of a scorpion.

Ireland's insect fauna

Insects appeared on earth about 400 million years ago and continued to evolve to exploit every conceivable terrestrial habitat. Their adaptability is virtually endless and their short reproductive cycle enables rapid evolution. They are highly mobile during their dispersal stage but the foldable wings in many species mean they can shelter in safety when the need arises. The most diverse groups are those that have complete metamorphosis which enables the adults and larvae to exploit different resources in a habitat. The intricate adaptations of insects to their environment is only partly understood for most species and we know virtually nothing other than a name and description for many species.

Relative to other plant and animal groups, Ireland's insect diversity is huge, accounting for over 11,400 species. Yet this is only about half the number insect species known to occur in Britain and about third of that known from France. Our cool, temperate climate and relatively isolated location on the northwestern fringe of the Europe limits the number of species that can colonise and survive here.

Ireland's insect fauna is dominated by two very large orders: the flies (order Diptera), and the bees, wasps, ants and other insects that make up the order Hymenoptera. Between them these two groups comprise over 20 per cent of all species of fauna and flora recorded from Ireland's terrestrial, aquatic and marine habitats.

The diversity of these two groups is enormous. Flies occupy a vast array of niches in terrestrial ecosystems. About a third of them have aquatic larvae and most are dependent on moist habitats for the larvae to develop. In fresh water they live at the bottom of the deepest lakes and the smallest ponds on the highest mountains.

The Hymenoptera are economically and ecologically very important insects. The vital role of bees as pollinators of plants is well known but there is a hidden diversity

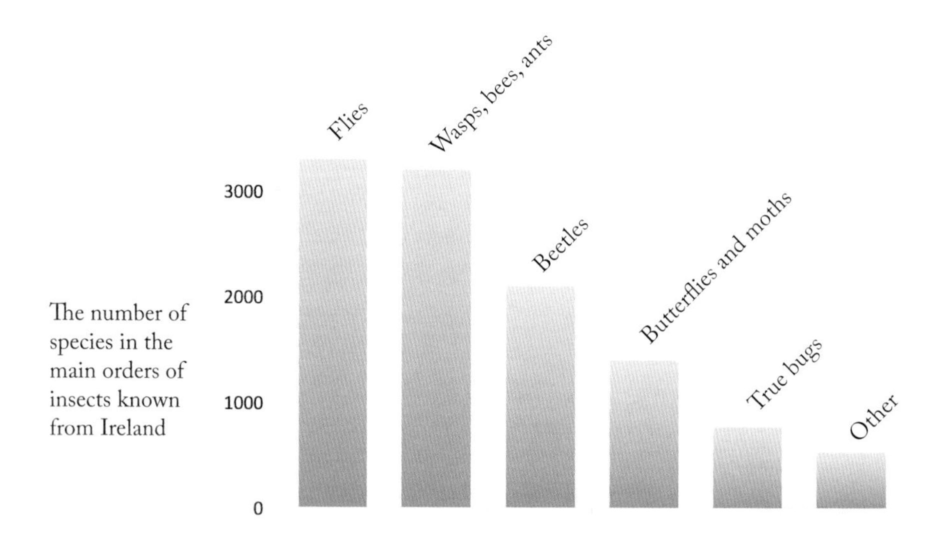

The number of species in the main orders of insects known from Ireland

amongst the vast numbers of minute predatory and parasitic wasps that feed and develop on other insects and invertebrates. Most hymenopterans require a nutritious diet and they generally feed in or on the bodies of living animals. A few have adapted to exploit the most nutritious parts of plants such as the gall-forming wasps that induce mutant growths on their host plant and juvenile bees that feed on protein-rich pollen collected by the workers.

At the other end of the scale there are less diverse orders of insects. The fleas, for example, have only 40 species represented in Ireland but their diversity is limited by the number of potential host species occurring here.

The conservation status in Ireland of a number of insect groups has been assessed in recent years, including bees, water beetles, butterflies, dragonflies and mayflies. These Red List assessments have found that up to one third of our insect species are under threat of extinction. This is as a result of a number of factors, including pollution, agricultural intensification and habitat loss. This is a worrying trend: not only a cause for concern for wildlife loss in general but also in terms of the essential ecosystem services that these small, inconspicuous creatures provide for us, including breakdown of waste products thus releasing nutrients back into soil. The first step to improving their conservation status is awareness. We hope that this book helps to improve awareness and appreciation of the value of Ireland's insects

The importance of insects

David Attenborough has famously said, 'If we and the rest of the back-boned animals were to disappear overnight, the rest of the world would get on pretty well. But if the invertebrates were to disappear, the world's ecosystems would collapse.'

Insects provide five major and essential ecosystem services: decomposition and nutrient recycling, herbivory (i.e. the eating of plants), a food source, seed dispersion and pollination.

By recycling dung, dead wood material (e.g. fallen tree trunks or dead leaves) or carcasses of dead animals, insects stimulate and accelerate breakdown of organic materials by such organisms as soil mites, fungi and bacteria, enhancing soil fertility and reducing the potential spread of diseases. Burrowing insects mix organic and inorganic material, increasing soil porosity and water-holding capacity.

Insects eating plants can cause substantial damage during their outbreaks. Yet it is known that most plants can compensate for being eaten, and usually the weaker trees/plants are eliminated. Besides, insects also produce very large amounts of nutrient-rich material (insect frass, moulted skins, bodies and partially eaten foliage) that falls as litter. Nutrients that leach from this litter stimulate turnover of organic material in the soil and enhance its health.

Many birds, mammals and reptiles feed on insects, which constitute a significant and necessary part of their protein diet. Major freshwater fisheries, especially those of salmonids, are supported largely or entirely by aquatic insects.

Insects disperse seeds, transmit pathogenic agents, and even transport other invertebrates from place to place. Ants are amongst the most important seed dispersers.

Many plants depend on insects as pollinators of their flowers, and pollinators can be a limiting resource for them. In Ireland, crops such as apples, strawberries, clover and oilseed rape all benefit from pollination and the value of this service to the economy has been estimated at €53 million per year.

How to find insects

Insects are found practically everywhere. Butterflies and bumblebees are easily seen feeding at flowers in gardens, and dragonflies at small ponds and lake edges. Grasshoppers are more often heard than seen on a warm summer's day in a meadow but can be disturbed and caught with a butterfly net. Shield bugs and ladybirds are more elusive and are most easily found by placing a sheet or umbrella underneath bushes and gently beating branches to dislodge them.

Some basic equipment can be very useful: a butterfly net is essential when learning about different butterfly, bumblebee, dragonfly and grasshopper species; a canvas or plastic sheet and stick for beating bushes are useful when searching for ladybirds and shield bugs; in addition, a hand lens can help with some of the finer identification features such as pollen baskets on bumblebees. Binoculars are also useful for identifying butterflies and dragonflies, which may be difficult to approach.

Ten tips for gardening for insects

1. Create a pond. Creating a pond is one of the best things you can do for wildlife. It will attract insects such as dragonflies to breed and many others. Lots of marginal plants and shady nooks will provide perfect conditions for all manner of insects.
2. Look after mature trees. A number of features of old trees make them extremely important for wildlife, including dead wood on the ground, water pools, holes and hollows inside the tree itself. If your garden is too small for big trees, try to get some planted in your local area and protect any that are there already.
3. Create a mini wildflower meadow. Meadow flowers will attract many different kinds of insects. A wildflower meadow does not have to cover your whole garden. You can create one on just a few square metres of soil.
4. Choose native. Choose a variety of native plants to attract a wider variety of species. Ivy, holly, honeysuckle, hawthorn, buckthorn, rowan and oak are just some of the many native species that are attractive in gardens and important for Irish insects.
5. Plant nectar plants for bees, hoverflies, moths and butterflies. Avoid plants with double or multi-petalled flowers. Such flowers may lack nectar and pollen, or insects may have difficulty in gaining access. Choose sunny, sheltered spots when planting.
6. Provide nectar throughout the spring, summer and autumn. Try to provide flowers right through the season. Spring flowers are vital for bumblebees and butterflies coming out of hibernation and autumn flowers help them build up their reserves for winter.
7. Avoid insecticides and pesticides. These kill many pollinating insects, as well as ladybirds, ground beetles and spiders.

8. Create a bug hotel. Dead and rotting wood is important for a number of invertebrates. Making a log pile in a shady spot can attract spectacular beetles if you are lucky, but the less glamorous characters are important too.
9. Avoid peat compost. Peat bogs are home to many special animals and plants, including the Large Heath butterfly, which is declining in Ireland. There are now good alternatives available from garden centres.
10. Relax! Don't feel that you have to be too tidy. Leave some areas undisturbed. Piles of leaves and twiggy debris in a hedge bottom or out-of-the-way corner will provide shelter for a range of insects and other creatures, and the seeds in dead flower heads can be valuable food.

Butterflies

Butterflies and moths are among the most beautiful, charismatic and fragile of all Irish insects. They belong to the order Lepidoptera, which translates as 'scaly wing'. It is a very diverse order with more than 157,000 known species, with at least as many again still unknown to scientists. Moths make up by far the greater part of the order. In Ireland, there are over 1,400 species of moths and only 35 species of butterflies.

'What's the difference between a butterfly and a moth?' is a common question, one that does not, however, have a simple answer. The division is artificial and the popular idea that butterflies are brightly coloured day-flying insects and moths are dull night-fliers is quickly dispelled by comparison of the Dingy Skipper butterfly or the day-flying, red-and-black 6-spot Burnet Moth. A convenient difference between Irish butterflies and moths is in the antennae (which are knobbed or clubbed on butterflies). Moths' antennae are various shapes, but only the burnet moths have clubbed antennae.

Dingy Skipper Butterfly

6-spot Burnet Moth

This chapter is an introduction to Irish butterflies and will help identify all 35 resident species and regular visitors, as well as providing some tips to identify their caterpillars. Butterflies are relatively easily identified, readily seen on sunny days, and have some fascinating behaviour, including nest-building, migrating over thousands of kilometres, complex courtships and even caterpillar singing!

Life history – how butterflies and moths develop

Butterflies and moths go through four life stages: egg, caterpillar (larva), chrysalis (pupa) and adult (imago). Eggs are usually laid on food plants and are fixed to the surface with special glue that hardens rapidly. The caterpillar's first meal is often its own eggshell before it begins to feed on the food plant. Some caterpillars spend practically all the time in search of or eating food. Many species eat only young, easily digested and relatively nutritious plant tissues and most can survive only on certain plant species. Caterpillars may also spend significant amounts of time basking in the sun in order to raise their body temperature, particularly species that are active early in the year and it is partly to facilitate digestion of their vegetable diet. Caterpillars are often cryptically coloured which helps them avoid detection by predators or parasitoids. This is especially true of the caterpillars of the Satyridae (browns) whose caterpillars are almost impossible to find amongst the grasses on which they feed.

When a caterpillar is fully grown, hormonal changes induce it to stop feeding and search for a suitably secluded pupation site, such as the underside of a leaf or branch. Caterpillars of the family Lycaenidae (hairstreaks, coppers and blues) attract ants with calls and offerings of tasty secretions. In exchange ants provide protection to the developing insect at a vulnerable stage. Most caterpillars anchor themselves firmly before the final moult and transformation into the chrysalis. The chrysalis is typically well camouflaged and difficult to find in the wild. Inside it an amazing transformation occurs as the adult butterfly takes shape. When the time is right, the butterfly pulls itself out of the chrysalis skin and within a few minutes the wings are fully expanded and the insect is ready for flight.

Four life stages of the Garden Tiger Moth: egg, caterpillar, chrysalis and adult.

Butterfly and caterpillar anatomy – naming the parts

A butterfly's most dramatic feature is its wings. The colours and patterns come from layers of tiny scales that protect the wings and provide insulation. Typically, the scales on the top of a butterfly's wings are brightly coloured, while those on the underside are cryptically coloured for camouflage.

The butterfly body has three basic parts: head, thorax, and abdomen. Along with its proboscis, the long, straw-like tube used for drinking, many of the butterfly's sensory organs are on the head. The compound eyes, moveable, segmented antennae and labial palps at the base of the mouthparts all help the butterfly decide what is and what is not food. They also have taste organs on the feet that help it sense the correct food plant, which is especially important for egg-laying females.

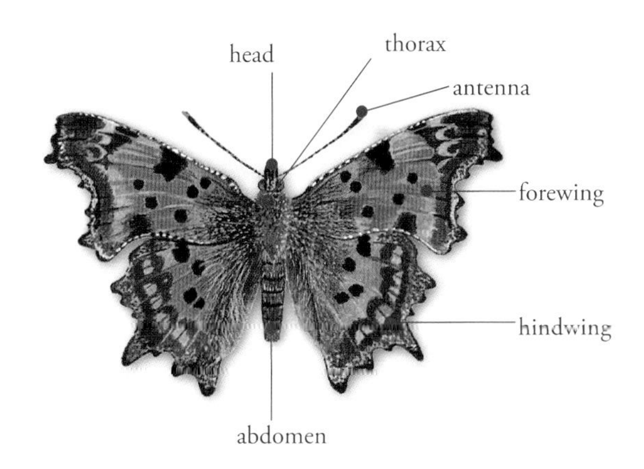

Caterpillars have what look like large compound eyes taking up most of the head. In fact, they only have simple light receptors and limited vision. They do, however, have other sensory receptors on the head and often wave the head around to assess information about their surroundings. There are three pairs of legs on the thoracic segments just behind the head. There are five pairs of prolegs on the abdominal segments. Many species that feed in exposed situations have dramatic adornments on the body such as branched spikes on the caterpillars of the Nymphalidae.

Peacock caterpillar

Large White on Buddleja

Large White caterpillar

Male and female
Brimstones on Buckthorn

Identification

Key features for butterfly identification are overall colour and patterning, time of year and habitat. With a little practice, most species can be identified while they are flying or feeding on flowers. Sometimes close examination is needed and the insects should be caught in a soft net and temporarily confined before being released. Most butterfly species can also be identified by their caterpillar stage, although for some species this is tricky as they may be confused with caterpillars of some of the 1,400 or so species of moth.

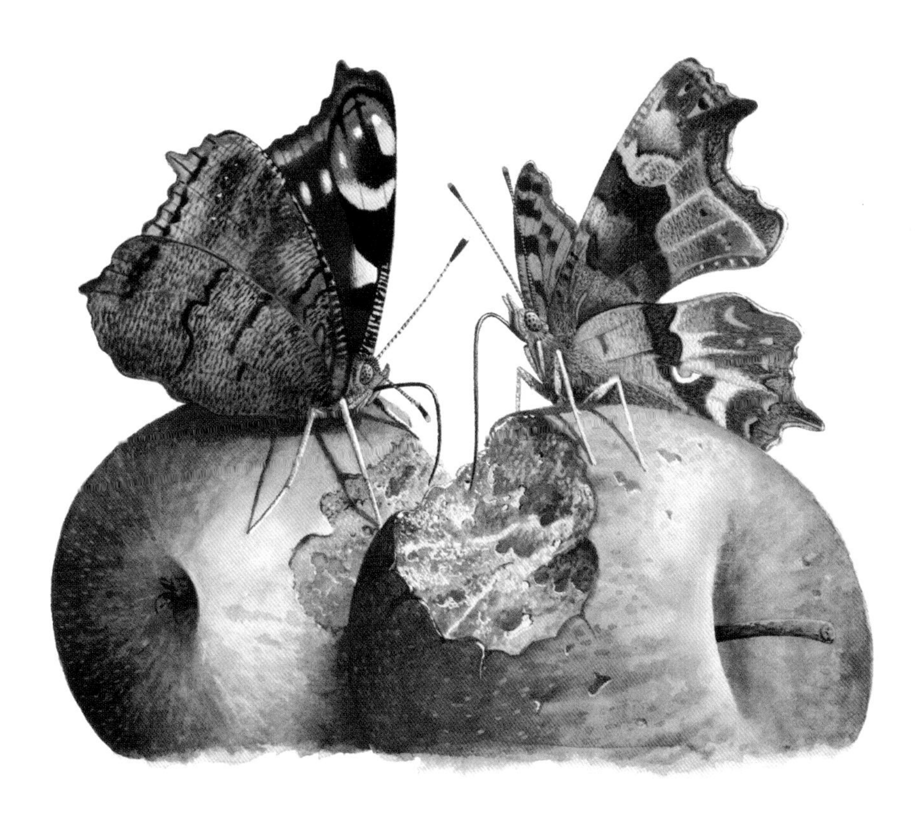

Peacock (left) and Comma feeding

Vanessids (five species)

The vanessids are the largest and most colourful butterflies in the Irish countryside and have noble-sounding names such as Red Admiral and Painted Lady. Fortunately, they are also some of the most common and they are frequent visitors to gardens. Together with the fritillaries, the vanessids make up the large family of Irish butterflies known as the Nymphalidae. As nymphalids, they have only two pairs of walking legs in the adult stage, the front pair having evolved to become brush-like appendages held below the head. They are all powerful fliers, a characteristic that facilitates migrations to Ireland from warmer climates. The vanessids overwinter as adults, but only the Small Tortoiseshell and Peacock find the Irish climate mild enough to survive the winter. The Painted Lady, Red Admiral and Comma are regular migrants arriving in Ireland in early summer. Many of the vanessid caterpillars feed on nettle, a good reason to leave a place in your garden for this plant.

Small Tortoiseshell (wingspan 50–56mm)

Peacock (wingspan 64–75mm)

Painted Lady (wingspan 58–74mm)

Red Admiral (wingspan 64–78mm)

Comma (wingspan 50–64mm)

Small Tortoiseshell *Aglais urticae* Ruán Beag

DESCRIPTION: *Adult*: Wingspan 50–56mm. Wings are orange with black-and-yellow patches and blue spots along the edges. Undersides are grey and black. Sexes are similar. *Caterpillar*: Black and yellow above, occasionally mostly black with a yellow stripe on each side. There are pairs of branched yellow spikes on each body segment.

SEASON: Adults emerge from hibernation on warm, sunny days in spring. The egg and caterpillar stages each last about four weeks. It spends most of its life as an adult.

NATURE NOTES: Very common throughout Ireland in all habitats and frequently in gardens. Eggs are laid in batches of 60–100 on the upper leaves of nettle. After hatching, caterpillars build a communal web near the top of the plant from which they emerge to bask and feed. As they grow, they move to new plants, building new webs and leaving a trail of webs, decorated with shed larval skins and frass.

CONFUSION: Worn and faded specimens could be mistaken for the Comma. Caterpillars are similar to those of other vanessids feeding on nettle.

Peacock *Inachis io* Péacóg

DESCRIPTION: *Adult*: Wingspan 64–75mm. Large and conspicuous with prominent eyespots on each wing. Underside of hindwings streaked with grey and black. *Caterpillar*: Black, speckled with tiny white spots. Pairs of branched spikes on each body segment and four pairs of orange-brown abdominal prolegs.
SEASON: One brood per year. Adults emerge from mid-July on and are long-lived, surviving through to the following breeding season. Eggs are usually laid during May and caterpillars develop in May and June. The chrysalis stage is usually around July.
NATURE NOTES: Widespread and common throughout Ireland in woods, hedges, parks and gardens. Eggs are laid in clusters of 300–500 on the underside of nettle leaves in sunny places. Caterpillars initially stay together in a tangled web of silk but as they mature they gradually disperse and are fully grown after a month or so. The chrysalis stage lasts about two weeks.
CONFUSION: Adults are unmistakable but caterpillars could be confused with those of other vanessids feeding on nettle.

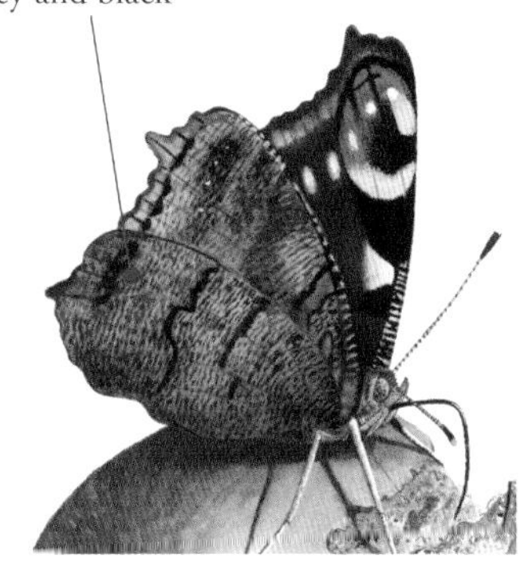

Painted Lady *Vanessa cardui* Áilleán

Description: *Adult*: Wingspan 58–74mm. Orange-and-black wings with black-and-white wing tips. *Caterpillar*: Mottled yellow-green and black with a yellow stripe along each side, becoming darker with age and with conspicuous pairs of branched spikes on each body segment.

Season: Adults are usually seen in Ireland between May and September. There may be several broods produced in a good summer.

Nature notes: This species migrates, over several generations each year, between northern Europe in summer and North Africa in winter. In some years it can be very abundant and arrives and breeds in Ireland in huge numbers. More usually there are modest numbers but it can be seen in most Irish habitats. Caterpillars feed on thistles, mallow and nettle. Eggs are laid singly on the upper leaves of plants. Caterpillars spend about four weeks feeding in silken tents, sheltering deeper within the plant after each moult. Pupation occurs low down in the food plant and the adult emerges after about two weeks.

Confusion: Adults are distinctive but caterpillars can be confused with those of other vanessids feeding on nettle.

Red Admiral *Vanessa atalanta* Aimiréal Dearg

DESCRIPTION: *Adult*: Wingspan 64–78mm. Distinctive bands of red and white on upper surfaces of the wings. Undersides are mottled brown and black. *Caterpillar*: Two colour forms occur: one is black with a yellow lateral stripe and the other yellow-green. Both have yellow, spiky bristles on each body segment.

SEASON: Migrants from Europe and North Africa arrive in Ireland, mainly during May and June but frequently from March onwards. Caterpillars may be seen from May through to late summer.

NATURE NOTES: Common in Ireland in all habitats. Adults feed on a wide variety of flowers, often opening and closing the wings conspicuously as they feed. Eggs are laid singly near the tip of nettle plants, often in shade. Caterpillars are solitary and feed within a tent of leaves held together with silk. When grown, they form a chrysalis inside the leaf tent. Recently, small numbers have been recorded overwintering in Ireland near the east coast.

CONFUSION: Adults are unmistakable. Caterpillars could be mistaken for those of other vanessids feeding on nettle.

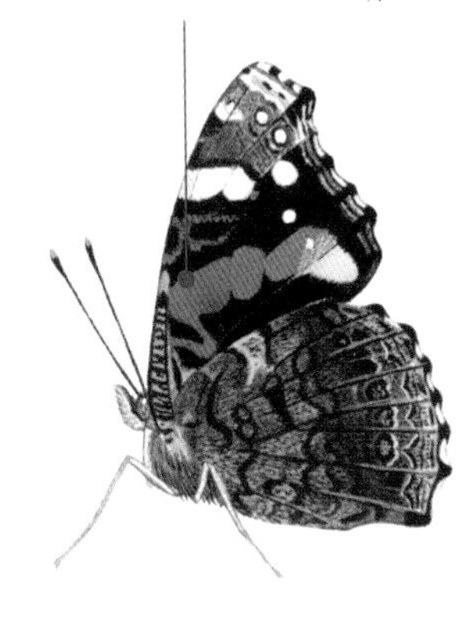

actual size

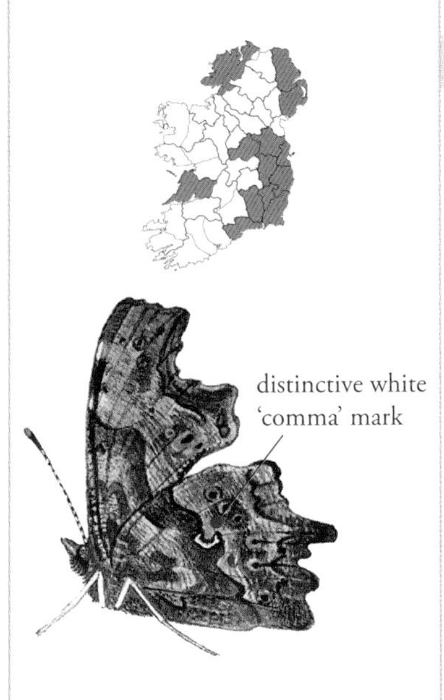

Comma *Polygonia c-album* Camóg

Description: *Adult*: Wingspan 50–64mm. Deep orange and brown on the upper side of wings with scalloped wing margins. The hindwings have a distinctive white 'comma' mark on the underside. *Caterpillar*: Black and orange with a large splash of white on the spiky back which resembles bird droppings.
Season: There are two broods per year. Overwintering adults re-emerge in April and a summer generation flies between June and August. A second generation appears in September and then hibernates. Caterpillars develop between May and August.
Nature notes: Previously an irregular migrant, it has now begun to breed in southeast Ireland. Adults frequent open woodland and hedgerows and are powerful fliers. Eggs laid singly on the upper side of fresh leaves of nettle usually in sheltered, sunny situations. Food plants also include elm, willow and currant. Caterpillars spin nests under leaves.
Confusion: Adults could be taken for old or tattered specimens of the Small Tortoiseshell.

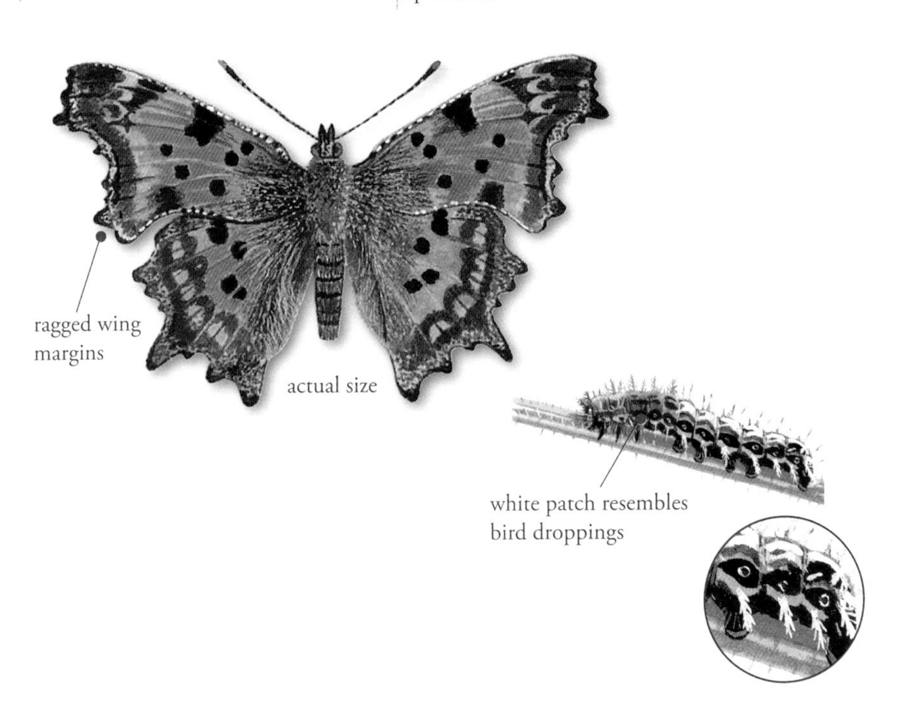

Fritillaries (four species)

The fritillaries are a subgroup of the Nymphalidae family and they are intricately patterned orange-and-black butterflies. The group includes our largest butterfly, the Silver-washed Fritillary. The four Irish species have intriguing life cycles and specific habitat requirements. Three species feed only on violets and require warm, sheltered yet sunny places for the caterpillars to develop. Caterpillars overwinter and become active early in spring to complete development and pupation. Caterpillars spend significant amounts of time basking during sunny spells. Three of the four species – the Pearl-bordered, Marsh and Dark Green Fritillaries – are on the Red List of Irish Butterflies and regarded as under threat of extinction in Ireland. The Marsh Fritillary is protected under the EU Habitats Directive because of substantial declines in its population across Europe.

Silver-washed Fritillary (wingspan 69–80mm)

Dark Green Fritillary (wingspan 58–68mm)

Marsh Fritillary (wingspan 42–47mm)

Pearl-bordered Fritillary (wingspan 44–47mm)

Silver-washed Fritillary *Argynnis paphia*
Fritileán Geal

DESCRIPTION: *Adult*: Wingspan 69–80mm. Males are golden-orange with black markings. Females are darker with more extensive black markings. Undersides of the hindwings are greenish with silver streaks. *Caterpillar*: Black with two yellow stripes along the back with brown, branched spikes on each segment.
SEASON: One brood per year. Adults fly between late June and early September. Caterpillars hatch in late summer, hibernate until spring and pupate in early June.
NATURE NOTES: Found throughout Ireland in deciduous woodland. Adults are powerful fliers and have a beautiful looping courtship flight. The female flies straight ahead while the male repeatedly loops under and over her, showering her with pheromones produced on the wings. Eggs are laid singly on tree trunks above suitably sheltered and sunlit patches of dog-violet. Eggs hatch after about two weeks in August and the tiny caterpillar immediately enters hibernation in a crevice in bark. In spring it descends to seek out fresh dog-violet leaves on which to feed and bask before pupation in late May and early June.
CONFUSION: Easily confused with the Dark Green Fritillary. The underside of the hindwings and the woodland habitat should confirm the identity.

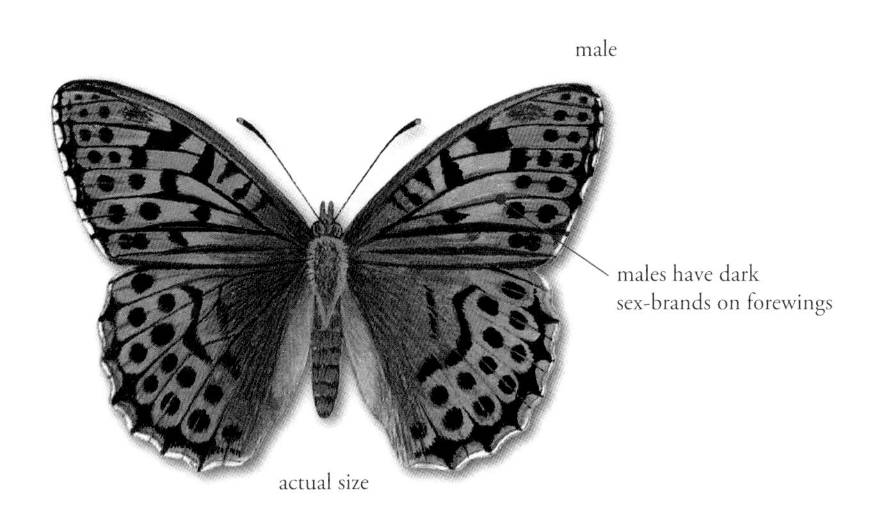

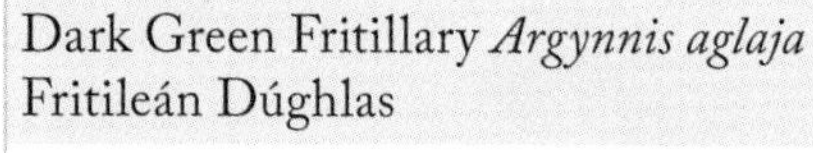

Dark Green Fritillary *Argynnis aglaja*
Fritileán Dúghlas

Description: *Adult*: Wingspan 58–68mm. Orange and black on the upper side of the wings. Undersides of hindwings are strongly marked with silver spots. *Caterpillar*: Black with a row of red, squarish spots along each side. Black, branched spikes on each body segment.

Season: Adults fly between June and August. Caterpillars overwinter before re-emerging in spring to complete development and pupation.

Nature notes: Widespread but not common, mainly in coastal locations. It lives in flower-rich habitats such as unimproved grassland, sand dunes and cutover bogs. Despite its powerful flight, it stays close to its breeding grounds. Eggs are laid on or near patches of violets in sheltered and sunny situations. Caterpillars hatch two or three weeks later and immediately hibernate in leaf litter for about nine months. In spring they feed and bask during sunny spells. The chrysalis is formed in tent of leaves and silk in May or June.

Confusion: Easily confused with the Silver-washed Fritillary but the undersides of the hindwings are clearly different, as is the habitat of each species. The Wall Brown is also orange on the upper side but it has small eyespots.

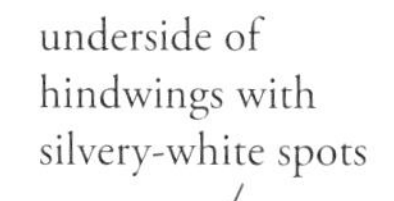

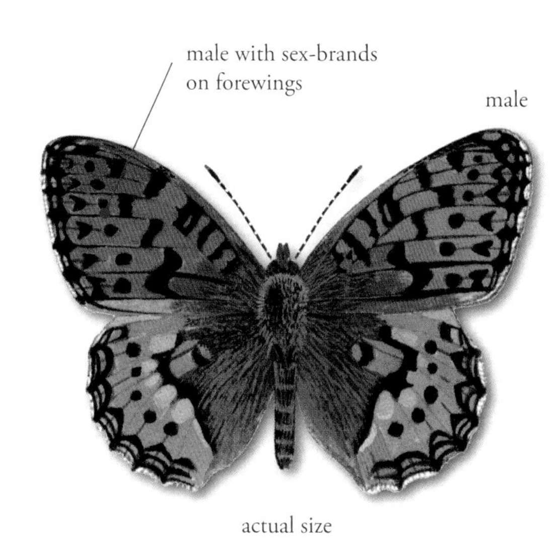

actual size

Marsh Fritillary
Euphydryas aurinia Fritileán Réisc

DESCRIPTION: *Adult*: Wingspan 42–47mm. Chequered orange, yellow and brown on the upper side, paler on the underside. *Caterpillar*: Black with white spots and with black, branched spikes on the body.

SEASON: Adults fly between mid-May and late June depending on weather. Caterpillars live from July to the following April and pupate in late April.

NATURE NOTES: Widespread but not common in Ireland in wet grassland, calcareous grassland, heaths and bog margins. Eggs are laid in clusters of 50–100, on the underside of leaves of the sole food plant, Devil's-bit Scabious. Only plants growing in sunny and sheltered places are used. About 30 days later caterpillars hatch and live in dense clusters in conspicuous silk webs. They move en masse to new food plants if necessary. In autumn they shelter low down in vegetation and hibernate, emerging as early as February to bask and feed as sunshine permits. Caterpillars become solitary when mature then pupate for two to four weeks in April or May depending on temperature.

CONFUSION: Possibly confused with the Pearl-bordered Fritillary in Galway and Clare but they are rarely on the wing at the same time.

chequered orange, yellow and brown

actual size

Pearl-bordered Fritillary
Boloria euphrosyne Fritileán Péarlach

Description: *Adult*: Wingspan 44–47mm. Sexes are similar. Adults are chequered orange and black above. Margins of undersides of hindwings have a row of pearl-like spots. *Caterpillar*: Black with a twin row of yellow spines.

Season: Adults fly for about six weeks from late April to mid-June. The caterpillar stage lasts from June to the following March.

Nature notes: Only known from scrubby grasslands and clearings in woods in Clare and southeast Galway and the Aran Islands. Eggs are laid singly on carefully selected violets in sheltered and sunny places and hatch after 11–20 days. Caterpillars feed and bask from June to September, then hibernate, half-grown, before resuming feeding in spring. Pupation takes place in April. The flight season varies with temperature with an extended pupation in cool weather.

Confusion: With the Silver-washed and Dark Green Fritillaries but much smaller and flying at a different time of year.

males are similar to females

actual size

Whites (eight species)

The white butterflies, or Pieridae family, are white and yellow and it may be that the original 'butter-coloured fly' was the Brimstone, a member of this family. Colouration varies slightly through the year; butterflies of the spring generation tend to be weakly marked while those of the summer generations tend to be more heavily marked. They hibernate as a chrysalis, with the exception of the Brimstone, which hibernates as an adult butterfly. This family feed primarily on plants in the cabbage and pea families, again with the exception of the Brimstone, which feeds on buckthorn and alder buckthorn. The Large White and Small White are regarded as pest species of cultivated cabbages and are closely associated with farmland and gardens.

Green-veined White (wingspan 40–52mm)

Small White (wingspan 38–57mm)

Orange Tip (wingspan 38–50mm)

Large White (wingspan 58–76mm)

male

female

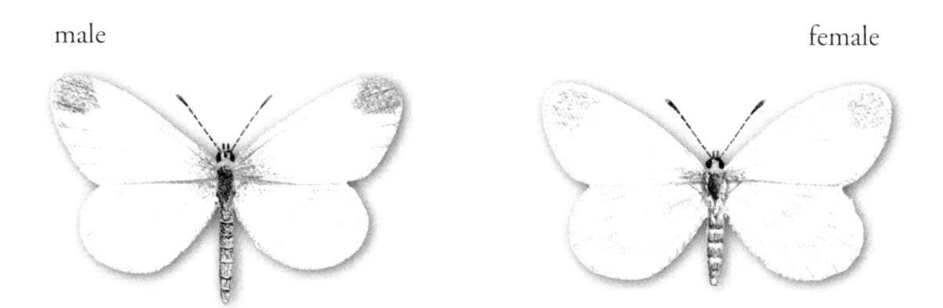

Cryptic Wood White (wingspan 42mm)

male

female

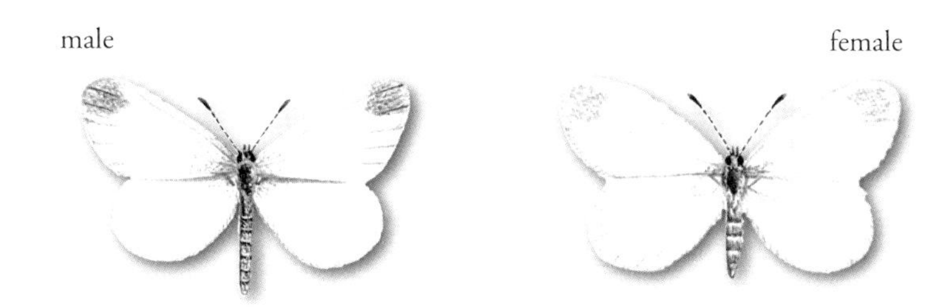

Wood White (wingspan 42mm)

male

female

Brimstone (wingspan 60–74mm)

male

female

Clouded Yellow (wingspan 54–62mm)

Green-veined White *Pieris napi* Bánóg Uaine

Description: *Adult*: Wingspan 40–52mm. Undersides of the hindwings are yellowish with distinct greenish veins. Individuals of second brood are more strongly marked. *Caterpillar*: Green body covered with small hairs and minute black dots.

Season: Adults of the first brood emerge from the chrysalis in April or May and fly until about June. Adults of the second brood fly during July and August and their offspring overwinter as a chrysalis.

Nature notes: The most common butterfly in Ireland, found in all habitats, even intensive farmland. Eggs are laid singly on the underside of leaves of food plants and hatch after about a week. Caterpillars feed on a range of wild crucifers including cuckooflower, watercress and hedge mustard. Caterpillars eat the eggshell before feeding on leaves and they develop within about four weeks. The chrysalis is generally formed away from the food plant, low down in vegetation and lasts around ten days unless overwintering.

Confusion: Easily confused with the other white butterflies in flight, in particular the Small White, but distinguished by the green veins on the wings.

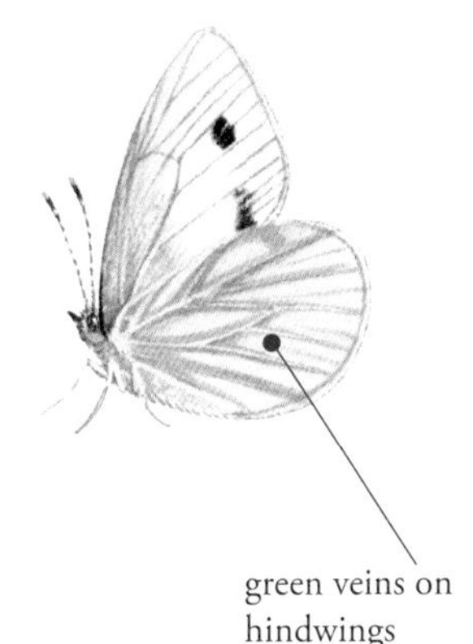

green veins on hindwings

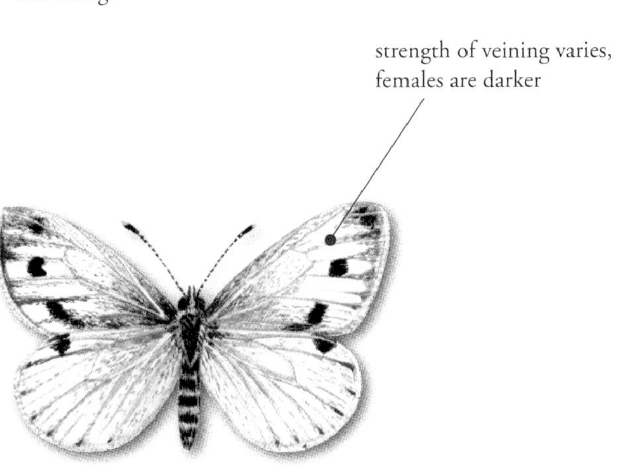

strength of veining varies, females are darker

actual size

Small White *Pieris rapae* Bánóg Bheag

Description: *Adult*: Wingspan 38–57mm. Wings are whitish with grey or black markings on the tips. Males have one spot on the forewing and females have two. The second brood has darker markings on the wings. Undersides of hindwings are pale yellow. *Caterpillar*: Bright green with minute black spots and fine, pale hairs. A thin yellowish line extends along the centre of the back.

Season: Two broods per year occur in Ireland. The first emerges from April to June and a second, larger emergence takes place between July and September. Caterpillars are common between June and September.

Nature notes: Found throughout Ireland in all habitats including gardens, arable crops and sand dunes. Eggs are laid singly on the underside of leaves. Caterpillars feed on a range of wild or cultivated plants including the cabbage family and nasturtiums in gardens and can be a serious pest.

Confusion: The Large White is larger with more prominent dark patches that extend down the outer edge of the wings. The Green-veined White has green veining on the underside of the hindwings and female Orange Tips have mottled green markings on the undersides of the hindwings.

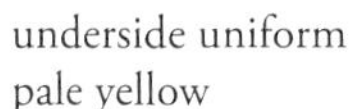

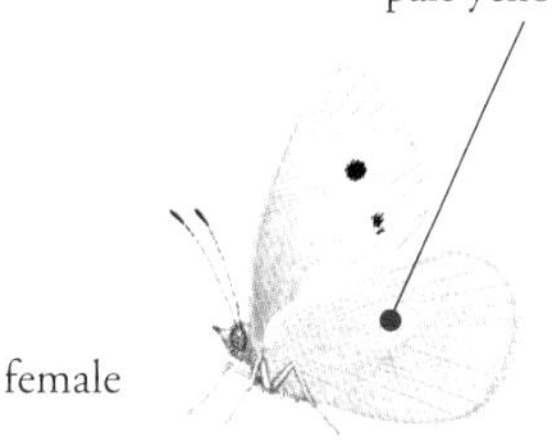

caterpillars often lie on the midrib of leaves

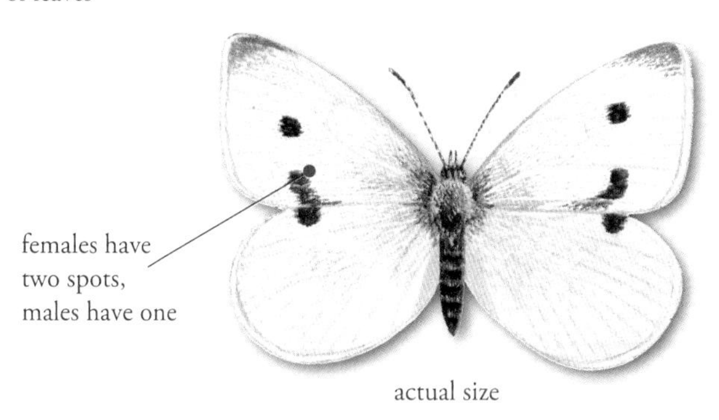

actual size

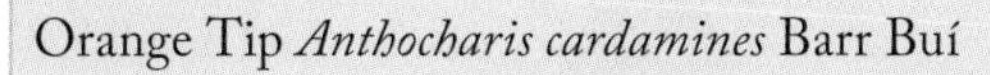

Orange Tip *Anthocharis cardamines* Barr Buí

Description: *Adult*: Wingspan 38–50mm. Males have a prominent orange patch on the forewing. Males and females have green marbling on the underside of the hindwings. *Caterpillar*: Pale green with a white stripe along each side.
Season: Adults fly between April and June. Caterpillars develop between April and June and the chrysalis lasts from June to the following April.
Nature notes: Very common throughout Ireland in damp pastures, hedges, roadsides and gardens. Males are easily noticed as they fly along hedges and roadsides, patrolling for females. Females are more sedentary, usually flying low down, investigating egg-laying sites. Eggs are laid singly near the flowering heads of crucifers including cuckooflower and garlic mustard and are quite conspicuous as they turn orange before hatching. Caterpillars feed on seedpods, and can be cannibalistic. When mature they seek dense cover before forming the chrysalis in which they rest until the following spring.
Confusion: Females could be confused with other whites, especially in flight, but the marbling on the hindwings is distinctive.

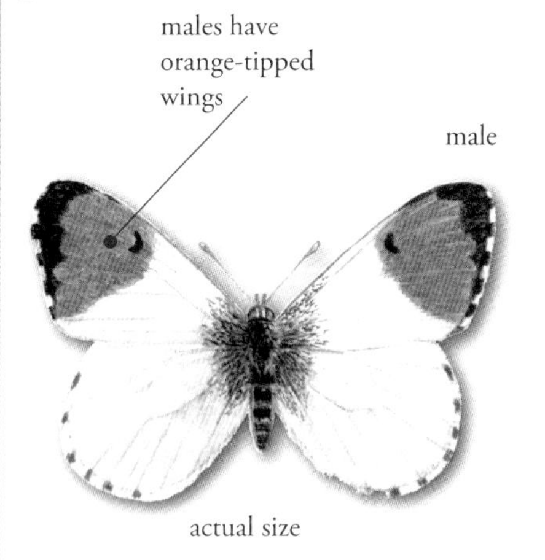

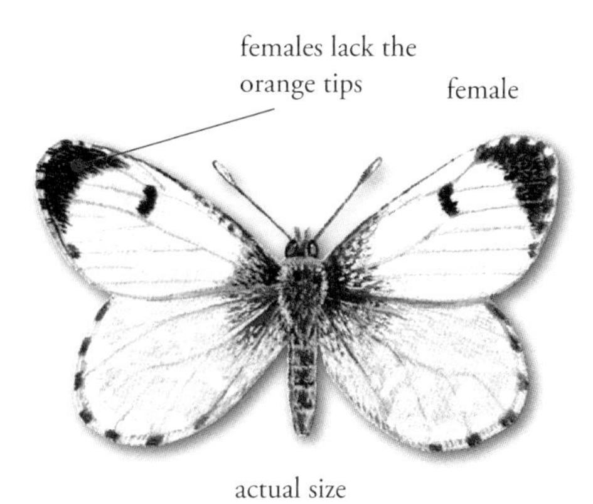

Large White *Pieris brassicae* Bánóg Mhór

DESCRIPTION: *Adult*: Wingspan 58–76mm. White with prominent black edges to the upper forewings and with two large black spots on the underside. Females also have spots on the upper side. *Caterpillar*: Up to 40mm and generally pale green with yellow stripes and black spots along the body.
SEASON: The first brood flies between April and June and the second brood flies between July and September.
NATURE NOTES: Widespread and common throughout Ireland in all habitats. It is a strong flier and can migrate large distances. Eggs are laid in clusters on plants of the cabbage family, especially cultivated varieties. Mature caterpillars seek suitable pupation sites on fences, tree trunks, and eaves of buildings. Pupation lasts around two weeks for the summer brood but around eight months for the overwintering brood.
CONFUSION: Small individuals can be confused with the Small White but the Large White is usually significantly larger and has more extensive dark marks on the wing tips.

both sexes have dark spots on underside of forewing

dark marks extend down side of wings

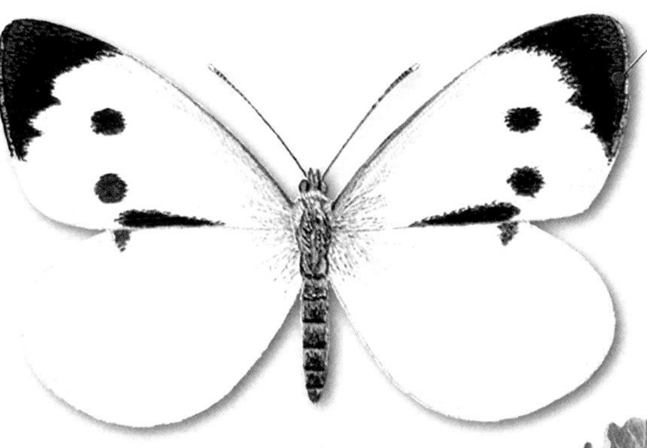

dark wing tips, grey in first brood, darker in second brood

actual size

Cryptic Wood White
Leptidea juvernica Bánóg Choille Dhuaithne

Description: *Adult*: Wingspan 42mm. Wings are white with light grey markings on veins and wing tips. Forewings of females are more rounded with paler wing tips. *Caterpillar*: Green with a yellow and dark green lateral stripe.
Season: There is one brood per year with adults flying in mid-May and early June.
Nature notes: Widespread and fairly common in sheltered open grassland, coastal dunes, unimproved meadow, grassy areas in open woods, road and rail verges and cutover bogs. Flight is very weak and males patrol with a floppy flight. Males court females by swaying and waving the proboscis and white-tipped antennae in front of potential mates. Eggs are laid singly on the food plants (bird's-foot-trefoil and meadow vetchling). Caterpillars develop during June and July and pupate in vegetation from August through the winter.
Confusion: The Small White is more robust and clearly tinged with yellow. The two Wood White species are indistinguishable in the field but have not been found together in Ireland. They were identified as separate species in Ireland in 2000 and the Cryptic Wood White does not occur in Britain.

wings are white with
pale grey markings

male

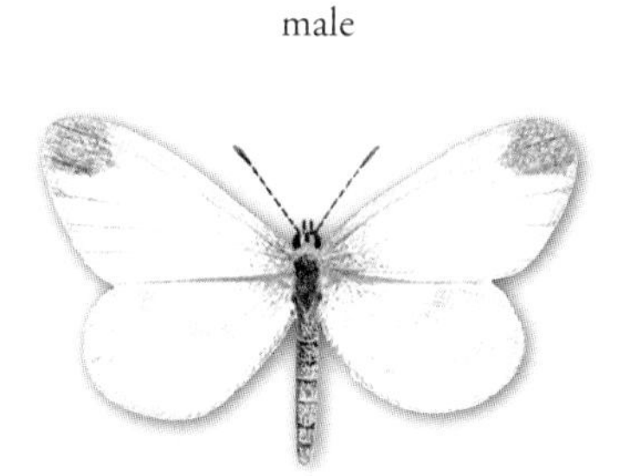

female

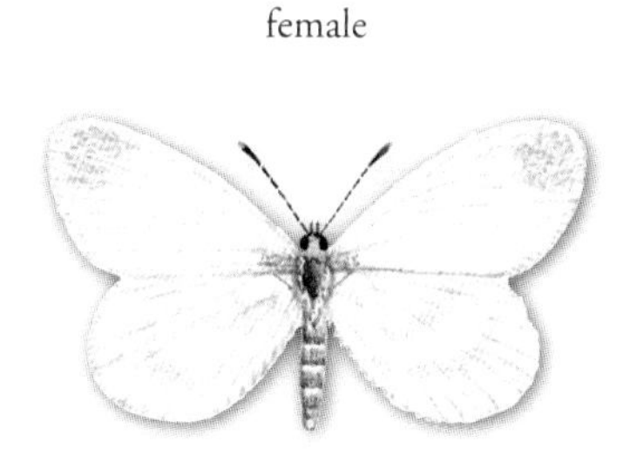

actual size

Wood White *Leptidea sinapis* Bánóg Choille

Description: *Adult*: Wingspan 42mm. Virtually identical to the preceding species and only distinguishable by examination of the genitalia under magnification. *Caterpillar*: Green with a yellow and dark green lateral stripe.
Season: There are two broods per year with adults flying in mid-May and early June and there is a smaller second brood in late July.
Nature notes: Confined to hazel scrub and open woods in the Burren and Galway. The combination of shelter and shade appear to be important for creating the right habitat for this species. Eggs are laid singly on the food plants (bird's-foot trefoil and meadow vetchling) in sheltered and sunny places. Caterpillars develop during June and July and pupate from August through the winter but some pupate earlier and emerge in late July.
Confusion: The Small White is more robust and clearly tinged with yellow. The Cryptic Wood White is indistinguishable in the field but these species have not been found together in Ireland. This species has only been found in areas of scrub and limestone grassland.

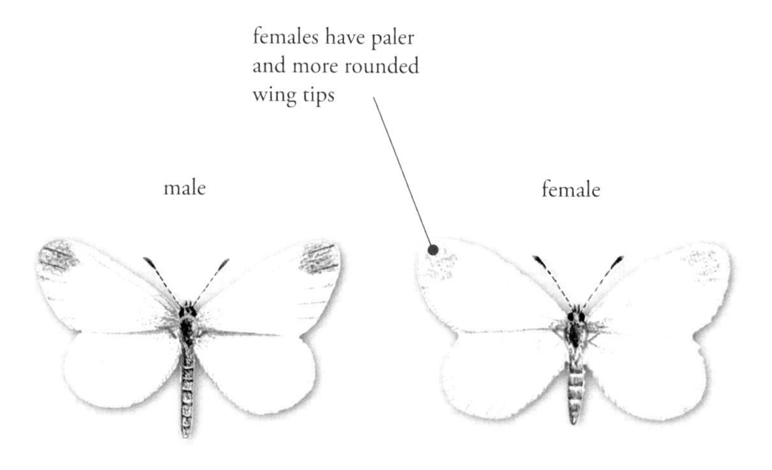

actual size

Brimstone *Gonepteryx rhamni* Buíóg Ruibheach

DESCRIPTION: *Adult*. Wingspan 60–74mm. Males are bright yellow and females are pale creamy yellow with small orange spots on each wing. Wings have sharply pointed margins, giving a leaf-like appearance. *Caterpillar*: Dull green with paler sides and white lateral stripe.

SEASON: One of the first butterflies of spring. Adults emerge from hibernation around March with males appearing first. Caterpillars develop between May and July and pupate in July and August. New adults appear from the end of July and hibernate from September.

NATURE NOTES: Widespread but not common, this butterfly is most abundant in limestone grassland, scrub and hedgerows in central and western Ireland where the caterpillar's food plants grow. Caterpillars feed only on buckthorn and alder buckthorn, both of which are sparsely distributed in areas with limestone. Adults are strong fliers and can occur in grassland and woodland edges some distance from the caterpillar food plants. Adults feed on thistles and other tubular flowers.

CONFUSION: Pale females in flight can be mistaken for the Large White.

Clouded Yellow *Colias crocea* Buíóg Chróch

Description: *Adult*: Wingspan 54–62 mm. Wings are orange-yellow with a thick black border on the upper side. Females have yellow spots within the black border. *Caterpillar*: Dull green with short, pale white hairs. There is a pale lateral line with red and yellow streaks within it.

Season: Migratory adults from southern Europe and North Africa arrive in spring or early summer. They may produce a brood in July or August but these die off during cooler weather in autumn.

Nature notes: Infrequently recorded throughout Ireland but most often near the south and east coasts. In most years numbers are low but occasional mass migrations occur. Eggs are laid singly on the flowers of clover and other legumes and caterpillars feed and grow rapidly. The second brood can be much more numerous than the migrants and most sightings are from late summer. Caterpillars may be able to survive mild winters in Ireland.

Confusion: The Brimstone is also yellow but with much less orange and no black on the wings.

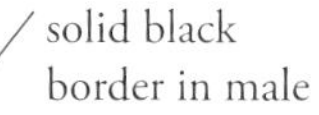

border with yellow spots in female

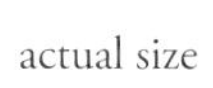

Blues (three), Hairstreaks (three) and Copper (one species)

The Lycaenidae family includes the blues, hairstreaks and Ireland's only copper. They are small with bright, metallic colours. They have intriguing, although poorly understood, relationships with ants. The caterpillar and chrysalis have a 'honey-gland' on their bodies, which secretes a nectar-like fluid to attract ants. In return the ants protect the caterpillar and chrysalis from predators and parasites. The caterpillar and chrysalis can also call to ants, making chirping noises by rubbing body segments together. The Small Blue has suffered substantial declines in population in recent years and is now considered as Endangered in Ireland on the Irish Red List.

Ants tending a lycaenid caterpillar

male

female

Common Blue (wingspan 30–38mm)

male

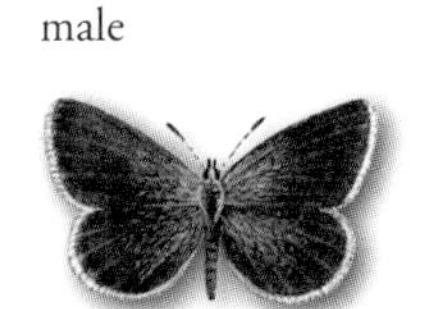

Small Blue (wingspan 16–26mm)

male

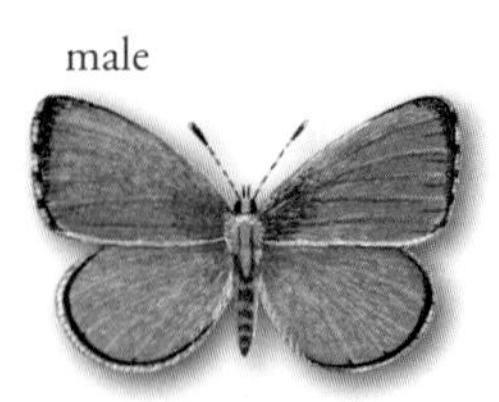

female

Holly Blue (wingspan 30–35mm)

male

Green Hairstreak (wingspan 27–33mm)

male

female

Brown Hairstreak (wingspan 36–42mm)

male

female

Purple Hairstreak (wingspan 32–39mm)

male

Small Copper (wingspan 32–35mm)

Common Blue
Polyommatus icarus Gormán Coiteann

Description: *Adult*: Wingspan 30–38mm. Males, blue above with orange spots on the undersides of hindwings. Females are blue and brown on the upper side with the bluest forms in northern and western areas. Females have orange spots on the upper and lower sides of each wing. *Caterpillar*: Bright green and furry with a small black head tucked into the body.
Season: In northern areas there is a single flight period in July. Elsewhere there are two flight periods in May/June and in August. Caterpillars overwinter low down in vegetation and pupate in spring.
Nature notes: Found throughout Ireland in sunny, sheltered areas with the food plant, bird's-foot trefoil, especially in dryish coastal areas. Eggs are laid singly onto the food plant and hatch after about nine days. The bright green caterpillar feeds by day and chirrups a tuneless song that is probably caused by the rhythmic compression of air in its abdomen. The chrysalis stage lasts about two weeks.
Confusion: Possibly confused with the Holly Blue but distinguished by the orange spots.

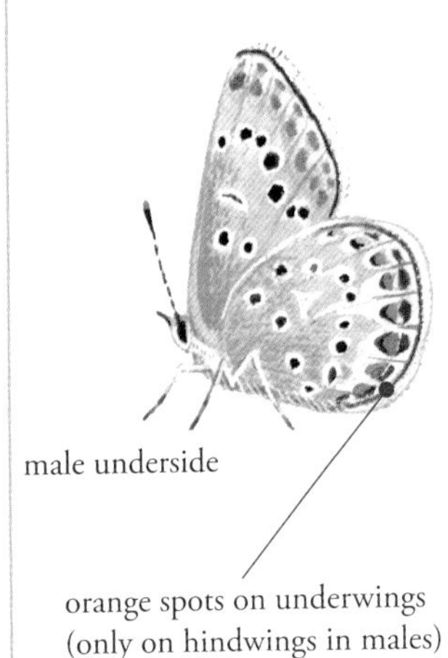

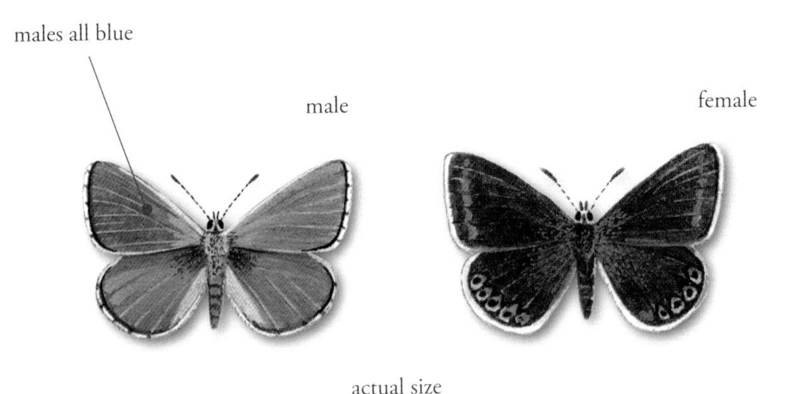

actual size

Holly Blue
Celastrina argiolus Gormán Cuillinn

DESCRIPTION: *Adult*: Wingspan 30–35mm. Males are blue above with fine chequered black marks near the tips of forewing. Females have black wing tips and in those of the second brood the outer third to half of the wing is black. Undersides of the hindwings are light blue with small black spots. *Caterpillar*: Most caterpillars are green but some have reddish markings.
SEASON: In northern areas there is one brood and adults fly during July. Elsewhere there are two flight periods, the first from April to May and a second during August. Offspring from the second brood overwinter as a chrysalis.
NATURE NOTES: Widely distributed in Ireland but scarce in northern and western areas. Occurs in open wooded areas, hedgerows and gardens. Often seen fluttering high around trees. Eggs are laid singly at the base of flower buds of holly in spring and ivy in late summer but other plants are also used. Eggs hatch after a fortnight and caterpillars feed on flower buds. The late summer generation hibernates as a chrysalis until emergence in spring.
CONFUSION: The Common Blue has orange spots on the underside of the wings.

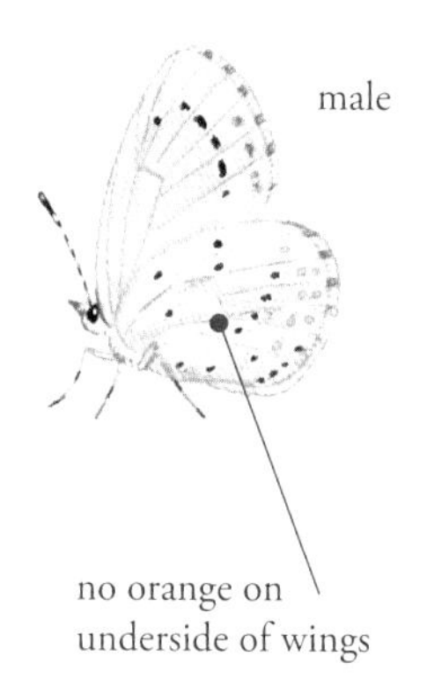

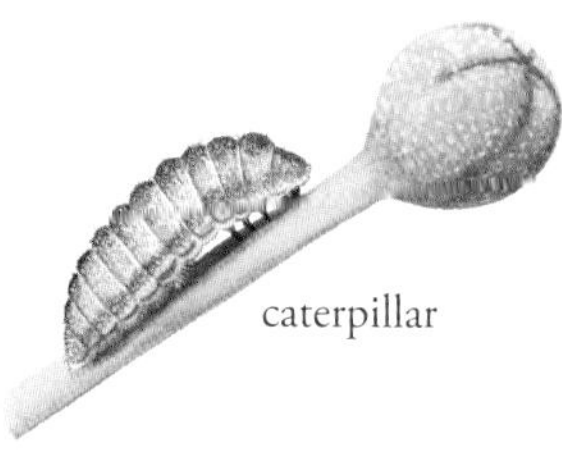

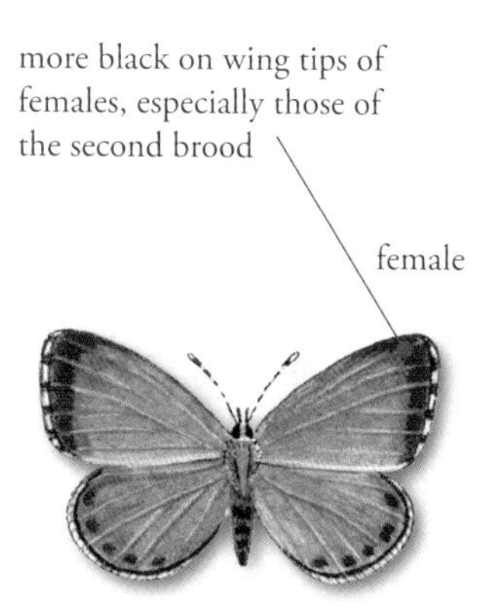

actual size

Small Blue *Cupido minimus* Gormán Beag

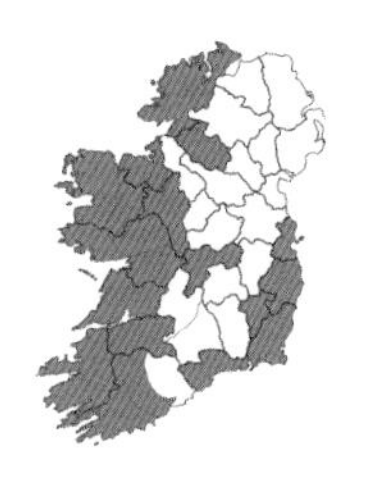

Description: *Adult*: Wingspan 16–26mm. Males are grey-brown on the upper surface and have only a scattering of blue scales on the wings. Females have little or no blue on the upper surface of the wings. Undersides are grey. *Caterpillar*: Pinkish-grey with a darker line down the back.

Season: Adults fly from late May to June. Caterpillars develop over about four weeks and by late July caterpillars enter hibernation until April or early May. Then they pupate and the adult butterfly appears after about two weeks in late May.

Nature notes: Widespread but very local in Ireland, on dry, sheltered grassland in coastal areas, quarries, eskers and the Burren. Adult females lay eggs singly onto flowers of the sole food plant, kidney vetch. The well-camouflaged caterpillars feed on developing seeds in flower heads. Fully grown caterpillars and chrysalises can produce 'songs' to attract ants, which help protect the defenceless pupa. Irish populations are severely fragmented and it has recently become extinct in Northern Ireland.

Confusion: Differs from the other two blue Irish butterflies due to its tiny size and colouring.

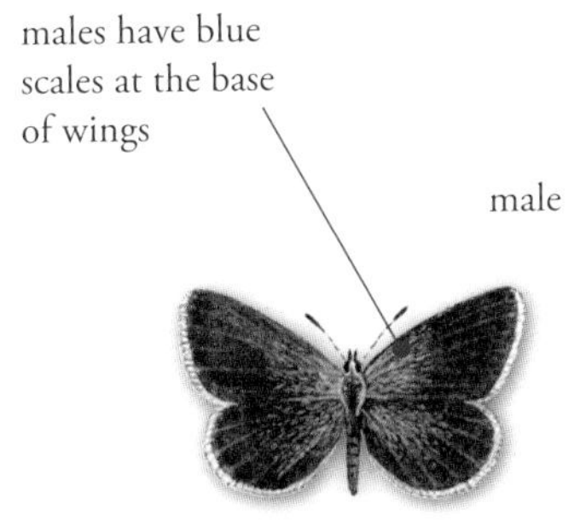

Green Hairstreak
Callophrys rubi Stiallach Uaine

Description: *Adult*: Wingspan 27–33mm. Sexes are similar. Wings are brown on the upper side and green underneath with a row of fine white spots on the hindwings. It always perches with its wings closed, showing the green underside and it can be difficult to spot. *Caterpillar*: Bright green and marked with oblique yellow stripes.

Season: Adults fly between late April and early July. Caterpillars develop during summer and mature around July. They pupate on the ground and remain in the chrysalis stage until the following spring.

Nature notes: Widespread in Ireland where there is gorse or bilberry on moorland, rough ground, bog margins and hillsides. Males are territorial and perch on prominent leaves and challenge intruders. Eggs are laid singly on the growing tips of gorse but other food plants are sometimes chosen including, bilberry, broom and bird's-foot trefoil. Caterpillars are sedentary and difficult to spot. The chrysalis can make a relatively loud noise and secretes substances attractive to ants, which guard the chrysalis.

Confusion: A distinct butterfly that is not easily confused with any other Irish species.

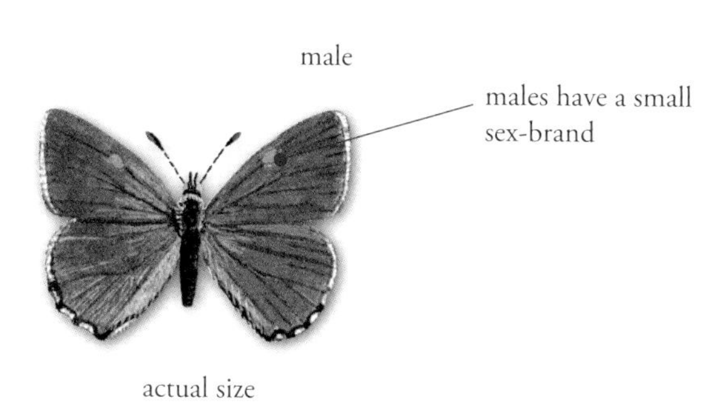

actual size

Brown Hairstreak
Thecla betulae Stiallach Donn

DESCRIPTION: *Adult*: Wingspan 36–42mm. Males are brown above with a small orange tail on the hindwings. The underside is orange with white streaks. Females have an orange band on the forewings. *Caterpillar*: Bright green and patterned with yellow oblique markings, turning purple when mature.
SEASON: There is one brood per year. Adults fly from mid-July to mid-September. Eggs laid in summer hatch the following May. Caterpillars grow during spring and pupate between June and July.
NATURE NOTES: Only found in the Burren, around Lough Corrib and east Galway in hedgerows and scrub with blackthorn. Males congregate in the canopy or on tall trees and are rarely seen. Females descend from the canopy to lay eggs singly on the outer branches of the food plant, blackthorn. The white eggs are placed at a junction or base of a thorn where they remain throughout winter. Caterpillars hatch in early May and feed hanging from the underside of fresh leaves. When mature they turn purple and pupate in leaf litter or soil. The chrysalis is attractive to ants, which protect the chrysalis.
CONFUSION: Unlike any other Irish species.

white streaks on orange underside

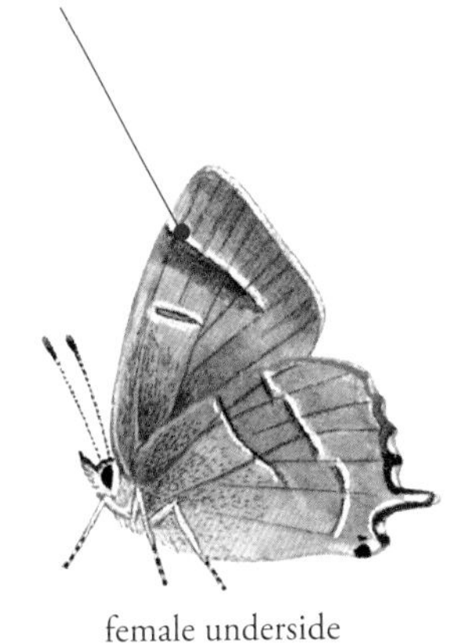

female underside

males have brown above with sex-brands

male

female with orange marks

female

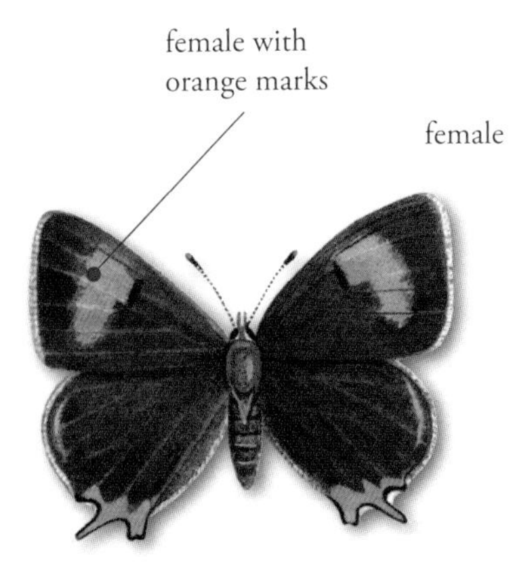

actual size

Purple Hairstreak
Quercusia quercus Stiallach Corcra

Description: *Adult*: Wingspan 32–39mm. Males are dark with a purple sheen above and grey below with white streak and an orange eyespot and tail. Females are darker with a reduced purple area on the forewing. *Caterpillar*: Brown and shaped like an elongated woodlouse.

Season: There is one brood per year. Adults fly from mid-July to early September. Eggs laid in summer remain dormant for eight months until April the following year. Caterpillars emerge in April and mature around June and pupate during July.

Nature notes: Widespread but not common in woodlands with mature oak trees. Perhaps more common than currently recognised, as it is an elusive butterfly spending most of its life in the canopy. Adults drink honeydew from leaves. Eggs are laid singly, just below buds of oaks in sheltered and sunny places. Caterpillars feed on and in developing buds of oak. When mature they drop to the ground and become a chrysalis. The chrysalis is attractive to ants and it is thought that the protection of ants is important for successful pupation.

Confusion: No other species has similar markings.

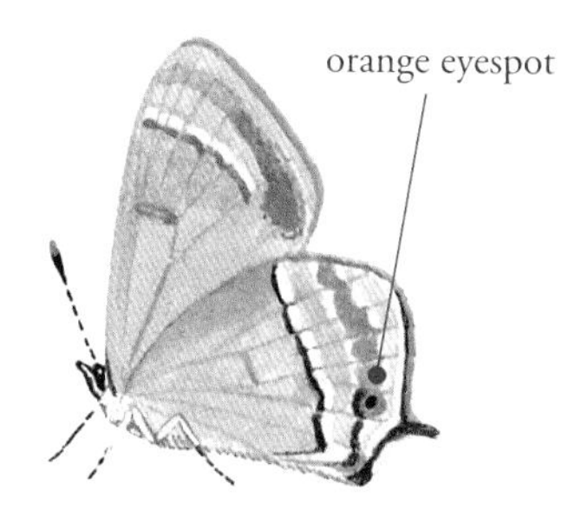

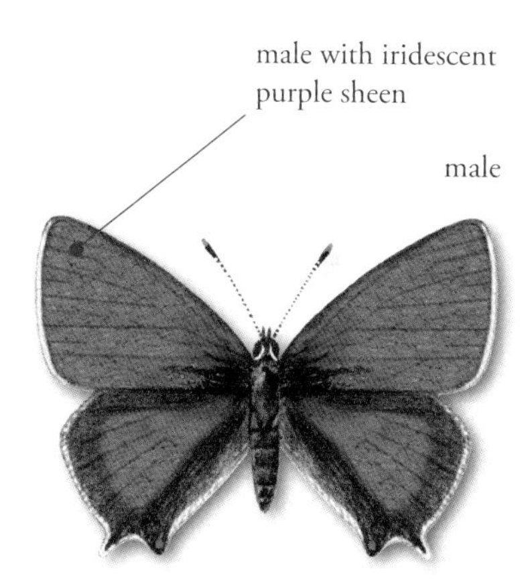

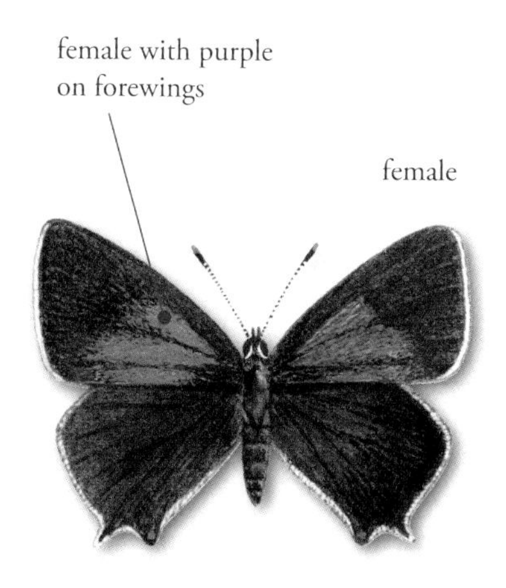

actual size

Small Copper *Lycaena phlaeas* Copróg Bheag

DESCRIPTION: *Adult*: Wingspan 32–35mm. Small with bright, iridescent, copper-coloured forewings with brown spots. Hindwings are brown above with a copper band and occasionally there is a row of fine blue spots. *Caterpillar*: Green and slug-like, sometimes with lines of purple on the body.

SEASON: There are two broods each year. Adults fly from late May to mid-June and from mid-July to August. Caterpillars from the second brood overwinter in vegetation from September until the following spring when they pupate.

NATURE NOTES: Widespread and common throughout Ireland in various habitats including grassland, woodland edges, heaths and road verges. It is an active butterfly: males defend territories from a prominent perch while females have a characteristic lazy flight when looking for egg-laying sites. Eggs are laid singly on the centre of sorrel and dock leaves. Caterpillars feed on the underside of leaves leaving distinctive translucent marks.

CONFUSION: Distinctive and not easily confused with any other Irish species.

greyish on underside

actual size

Browns (eight species)

The brown butterflies, or Satyridae family, are brown and orange with eyespots on the wings, which are thought to confuse predators. Caterpillars eat grasses and all the Irish species overwinter as caterpillars, perhaps feeding during mild weather. The Speckled Wood can also overwinter as a chrysalis. Four species are very common and can be found in grassy habitats all over Ireland, namely the Speckled Wood, Ringlet, Meadow Brown and Small Heath. The Gatekeeper is restricted to the southern and eastern counties while the Wall Brown and Grayling are restricted to warm dryish habitats and are most frequently seen near coasts. The Large Heath is regarded as Vulnerable to extinction in Ireland because of ongoing loss of its peatland habitat and severe declines in populations in recent decades. It is also regarded as Vulnerable on the European Red List because of population declines across Europe.

Speckled Wood (wingspan 47–56mm)

Ringlet (wingspan 42–52mm)

male

female

Meadow Brown (wingspan 52–56mm)

male

Gatekeeper (wingspan 44–48mm)

Grayling (wingspan 56–61mm)

Wall Brown (wingspan 45–53mm)

male

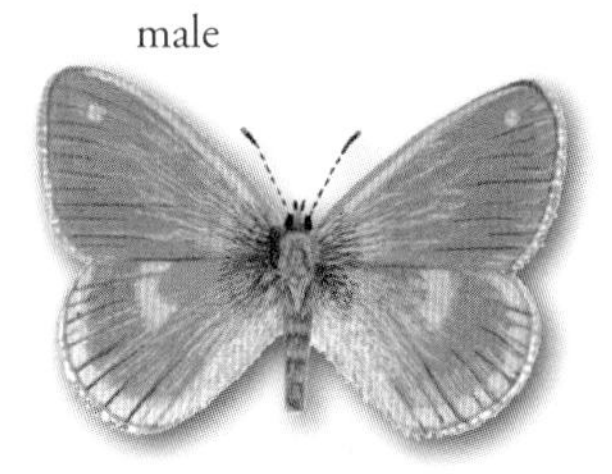

female

Large Heath (wingspan 40mm)

Small Heath (33–37mm)

cream-speckled wings

actual size

Speckled Wood
Pararge aegeria Breacfhéilachán Coille

Description: *Adult*: Wingspan 47–56mm. Brown with cream patches on the wings and with a row of eyespots on the hindwings. *Caterpillar*: The body is green with pale green-and-yellow stripes along the back and with rows of fine, white-haired tubercles. The tail end has two pale green points extending from each side.
Season: Flight season extends from April to early October with two or three overlapping broods each year. It can overwinter as a caterpillar or chrysalis and the first adults to emerge in April are those that overwintered as a chrysalis. These are followed, a few weeks later, by individuals that spent the winter as caterpillars. The flight periods of both broods and their progeny overlap throughout the summer.
Nature notes: Widespread and common in woodland, scrub and along hedgerows. Males perch in sunlit places and patrol along hedges and clearings and challenge other butterflies. Eggs are laid singly on a wide variety of coarse grasses including cock's foot and Yorkshire fog. In spring eggs are laid on grass in sunny places while summer eggs are laid in shady places.
Confusion: The cream speckled appearance is unlike other brown species.

female is dark brown

actual size

Ringlet *Aphantopus hyperantus* Fáinneog

Description: *Adult*: Wingspan 42–52mm. Males are dark brown to almost black, while females are dark brown. Both have a distinctive ring pattern on the wings that is clearer on the underside. *Caterpillar*: Light brown, covered in white hair with a dark stripe down the back that fades toward the front.
Season: Adults fly from the end of May to early September. Caterpillars develop over nine or ten months from August to the following May.
Nature notes: Widespread and common in Ireland in a variety of habitats including damp meadows, grassland and hedgerows. It can fly in cool weather and even light rain. Eggs released in flight or squirted onto grass when perched. Caterpillars feed on grasses in tussocks and are mostly dormant during winter but feed at night during mild weather.
Confusion: Possibly confused in flight with dark males of the Meadow Brown.

Meadow Brown

Maniola jurtina Donnóg an Fhéir

Description: *Adult*: 52–56mm. Males are brown with vaguely orange patches on forewings. The eyespot generally has one white dot in the centre. Females have more orange on the forewings. Hindwings are dull brown. *Caterpillar*: Bright green with fine pale lateral line and covered in pale hair.

Season: Adults appear around the second week in July and fly until mid-September.

Nature notes: Widespread and very common in meadows all over Ireland. Has a distinctive bobbing flight, at grass height, around knapweed or thistles. It also flies in dull weather. Eggs are laid singly on grass or sometimes released in flight and hatch after three weeks. Caterpillars partly develop, feeding on grasses, then overwinter and complete growth in spring. Pupation is in early summer and mass emergence occurs in July.

Confusion: Bright females can look similar to Gatekeeper but Gatekeeper has two white spots. Dark males could be mistaken for the Ringlet.

Gatekeeper *Pyronia tithonus* Geatóir

Description: *Adult*: Wingspan 44–48mm. Upperside of wings are orange with brown margins. The undersides of hindwings have small white spots. Males have a broad sex-brand on the forewing. There are two white dots in the eyespot. *Caterpillar*: Green or light brown with paler stripes along the sides of the body.
Season: Adults fly between late July and early September. Caterpillars hatch in summer then overwinter between October and March. They then complete growth and pupate around June.
Nature notes: Only found near southern coasts in scrub, hedgerow and open woodland. Males patrol along hedges and adults bask frequently. Eggs are laid in flight, ejected onto patches of grass near shrubs. Caterpillars feed on grasses and may be bright green or brown.
Confusion: Resembles the female Meadow Brown but it is smaller, the wings are more orange and there are two white dots in the eyespot.

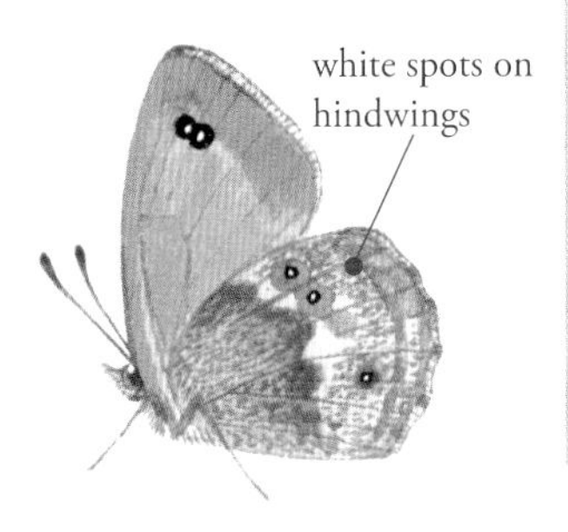

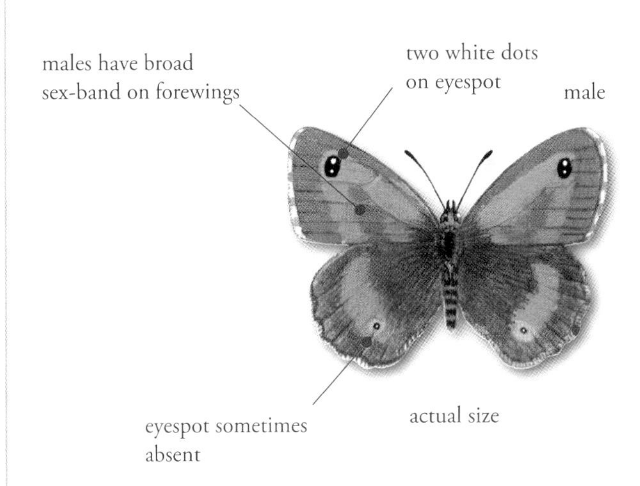

Grayling *Hipparchia semele* Glasán

Description: *Adult*: Wingspan 56–61mm. Upperside is mostly brown, washed with pale brown and orange. Undersides of wings are mottled grey and black. *Caterpillar*: Pale brown with dark brown and yellow stripes.

Season: Adults fly between late June and early September. Caterpillars live from late summer to the following May or June and then pupate.

Nature notes: Widespread across Ireland but mainly in dry, rocky places such as cliffs, quarries and sand dunes; most frequent on coasts. Also common in the Burren where the undersides of the wings closely match the grey limestone. Adults bask with wings closed on bare ground and are well camouflaged. Eggs are laid singly on fine grasses including fescues and marram. Caterpillars partly develop before hibernation. They resume feeding in spring and complete development and pupate underground around June.

Confusion: Meadow Brown is smaller and darker with the undersides more plain brown. The Gatekeeper is smaller and more orange on the upper side.

Burren form is grey rather than brown

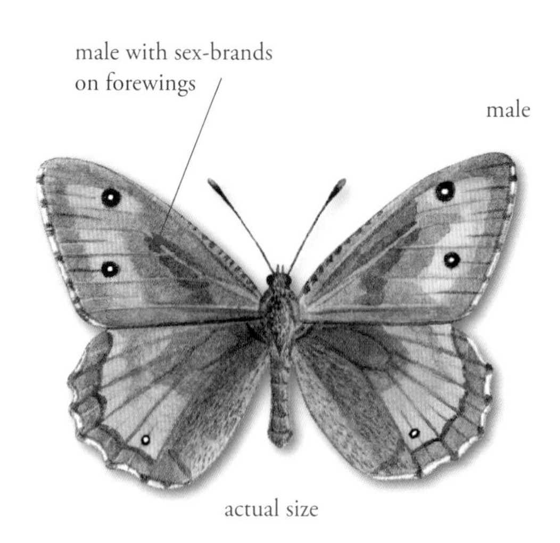

Wall Brown
Lasiommata megera Donnóg an Bhalla

Description: *Adult*: Wingspan 45–53mm. Adults are marbled orange and brown with small white eyespots on both pairs of wings. Undersides of hindwings are mottled beige and grey brown. *Caterpillar*: Head is green and globular. The green body is covered in fine white hairs with a pale stripe on each side.

Season: There are two broods each year. The first flies in May and June and the second, larger brood flies in August and September. Caterpillars and sometimes chrysalises of the second brood overwinter.

Nature notes: Found throughout Ireland but mostly in coastal areas and scarce in the north. It lives in grassland with patches of bare ground where soil is broken, exposing roots such as edges of paths or sheep scraps. It basks on walls or bare ground and occasionally flies long distances cross-country. Males patrol hedges and paths and pursue females or challenge males. Eggs are laid singly at the base or exposed roots of coarse grasses including cock's foot and Yorkshire fog.

Confusion: Gatekeeper, Grayling and the fritillaries have flashes of orange on brown wings but the combination of eyespots and mottled orange and brown is unique to the Wall Brown.

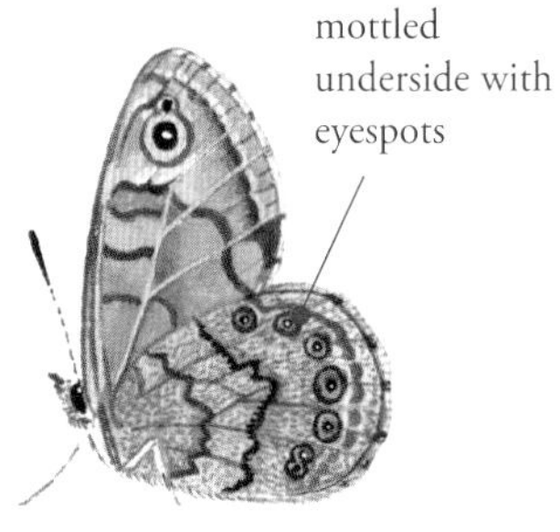

female

actual size

Large Heath
Coenonympha tullia Fraochán Mór

Description: *Adult*: Wingspan 35–40mm. Upperside of wings varies from bright orange-brown to dull orange. Hindwings are darker usually with small eyespots. Underside of the forewing has a pale streak near the eyespot. *Caterpillar*: Dark green with yellow longitudinal stripes and a large globular head that is wider than the body.
Season: There is one brood per year. Adults fly from early June to mid-July.
Nature notes: Widely distributed but not common in Ireland and absent from the eastern and southern coastal counties. The habitat is wet raised bog and blanket bogs, which are more prevalent in the north and west. Adults fly low over bog vegetation. Eggs are laid at the base of tussocks of hare's-tail cottongrass and hatch two weeks later. Caterpillars feed on growing tips of the food plant and enter hibernation around October. In spring they complete development and pupate around the end of May.
Confusion: Easily confused with the Small Heath with which it sometimes occurs on bog margins. The pale streak on the underside of the forewing is only found on the Large Heath.

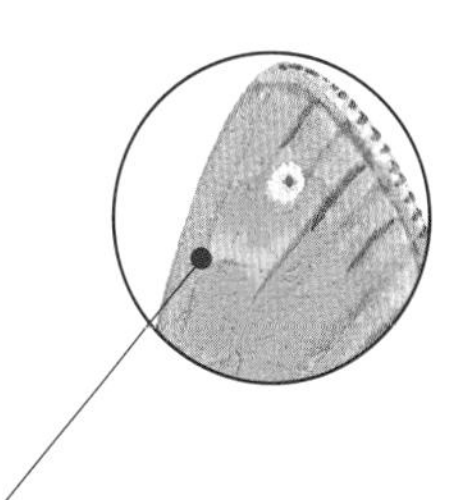

pale streak on underside of forewing

actual size

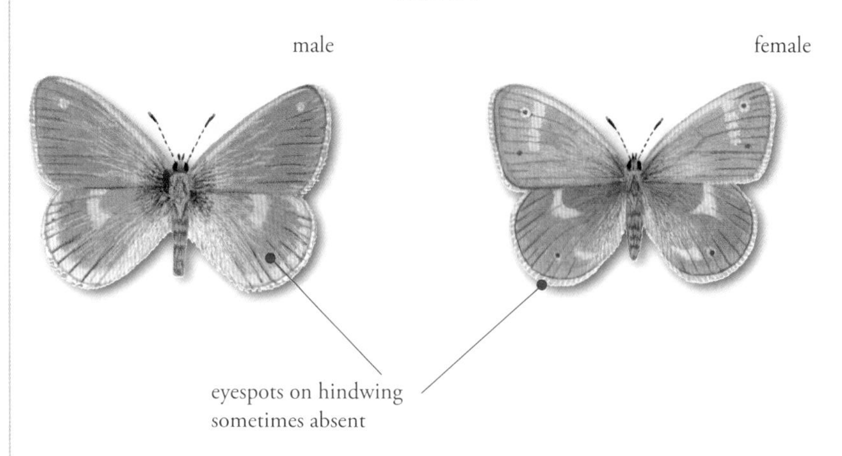

Small Heath
Coenonympha pamphilus Fraochán Beag

DESCRIPTION: *Adult*: 33–37mm. Upper surface dull orange to tawny yellow. There are no eyespots on the underside of hindwings. *Caterpillar*: Bright green with darker green stripes along the back and a white stripe on each side. The head is large and globular and wider than body.
SEASON: Adults fly between mid-May and June and there is a smaller second brood between August and mid-September.
NATURE NOTES: Widespread throughout Ireland but only most frequent in dryish places in unimproved grassland, heaths and dunes. Adults always perch with wings closed. Eggs are laid singly on grass blades and hatch after two weeks. Caterpillars feed low down in tussocks of fescues and meadow grass until late summer and then hibernate. They re-emerge in spring to complete development and pupate, suspended from a grass stem, around May.
CONFUSION: Large Heath is slightly larger and has a pale streak on underside of forewing, lives in wetter habitats and usually has eyespots.

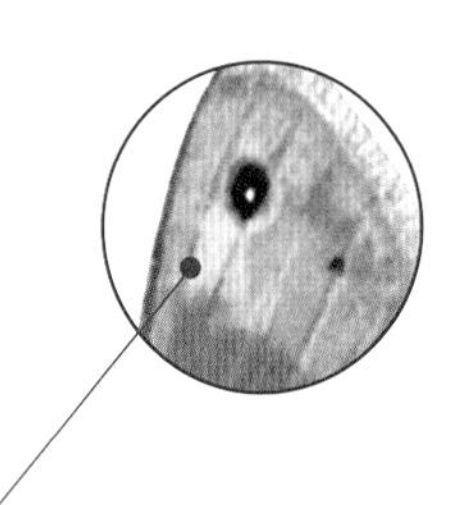

no pale streak on underside of forewing

actual size

Skippers (three species)

The skippers, or family Hesperiidae, are the most primitive of Irish butterfly families. They get their name from their low, fast, direct flight, almost 'skipping' along. They are small and resemble moths, folding their wings in a moth-like manner. Unlike other butterflies, the caterpillars spin silken cocoons in which they shelter. Relative to other Irish butterflies, they have large heads, well-separated antennae and short, dumpy wings. There are three Irish species, two of which are recently established in Ireland.

Dingy Skipper (wingspan 27–34mm)

Essex Skipper (wingspan 26–30mm)

Small Skipper (wingspan 27–34mm)

Dingy Skipper *Erynnis tages* Donnán

DESCRIPTION: *Adult*: Wingspan 27–34mm. Dull brown and moth-like, with mottled grey patches on the forewings. *Caterpillar*: The light green body is fattest in the middle and the head is black. There are pale stripes on each side.

SEASON: Adults fly between late April and late June.

NATURE NOTES: Absent from northeastern and southwestern counties. It is found in sheltered, warm places with bare ground such as limestone grassland, cutover bog, quarries and eskers but only infrequently on the coast. It sometimes occurs in damp places where it feeds on greater bird's-foot-trefoil. It is a fast flier but flies only short distances before stopping. Males seek out females and eggs are laid singly on the upper surface of young leaves of bird's-foot-trefoil. Eggs hatch after two weeks and caterpillars feed within a nest of silk. They are fully grown by August and spin a hibernation chamber. They re-emerge in early spring and pupate during April.

CONFUSION: Similar to some types of day-flying moths.

actual size

Essex Skipper
Thymelicus lineola Scipeálaí Essex

DESCRIPTION: *Adult*: Wingspan 26–30mm. A small orange-brown, moth-like butterfly with wings fringed with brown and cream scales. The undersides of the tips of the antennae are black. *Caterpillar*: Bright green with pale longitudinal stripes and a pale green head with brown-striped eyes.
SEASON: Adults fly between early July and mid-August.
NATURE NOTES: First found in Ireland in 2006 in County Wexford and subsequently in County Kildare. It lives in rough grassland and it is possible that eggs were accidentally imported in hay. Eggs are laid singly in grass sheaths, mostly of cock's foot, where they remain dormant for about eight months. They hatch around April; caterpillars feed in grass sheaths until about June, followed by pupation and emergence.
CONFUSION: Very similar to the Small Skipper. They can be distinguished only by the colour of the underside of the tips of the antennae: in the Essex Skipper these are black and in the Small Skipper they are orange.

actual size

Small Skipper
Thymelicus sylvestris Scipeálaí Beag

DESCRIPTION: *Adult*: Wingspan 27–34mm. Very similar to the Essex Skipper, this small orange butterfly looks like a moth. Females are slightly larger than males. Undersides of the tips of the antennae are orange. *Caterpillar*: Green with pale longitudinal stripes. The head and eyes are plain green.
SEASON: Adults fly from mid-July to mid-August.
NATURE NOTES: Discovered in County Kildare in 2005. It lives on ungrazed but not overgrown grassland. Like the Essex Skipper, it is likely that it was accidentally imported into Ireland in hay. Eggs are laid in small batches in sheaths of grass stems, especially Yorkshire fog, its preferred food plant. Eggs hatch after 2–3 weeks and the caterpillars immediately spin a silk tube in which they hibernate until the following April. They re-emerge and begin to feed on Yorkshire fog, hidden in tubes of sheaths and leaves until pupation.
CONFUSION: Very similar to the Essex Skipper. Distinguished by the colour of the underside of the tips of the antennae. In the Essex Skipper these are black and in the Small Skipper they are orange.

actual size

Ladybirds

Ladybirds are beetles belonging to the family Coccinellidae. There are 19 species recorded from Ireland out of well over 2,000 species of beetles. Fortunately, ladybirds are usually brightly coloured and well marked with spots and therefore relatively easily identified. A few species, however, are very variable and can cause confusion. The larvae of ladybirds are also distinctive enough to allow positive identification in most cases.

The name ladybird is thought to refer to 'Our Lady' as the colour red was associated with Jesus' mother, Mary, and the seven spots of the familiar 7-spot Ladybird correspond to the seven joys and sorrows. The Irish and Spanish names – '*boín Dé*' and '*vaquilla de dios*' (which both translate as 'God's little cow') – also have religious connotations.

Life cycle example: 7-spot Ladybird

1. Eggs (usually yellowish and elongate) are laid in batches of 20-40.

2. Larva are grub-like, often marked with spots of yellow, white and orange.

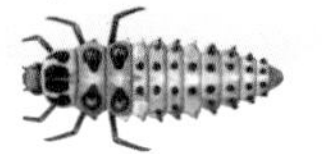

3. Pupae lack a hard coat but have an alarm reflex: when disturbed it flicks its body into an upright position in a startling fashion.

4. The adult beetle emerges after about two weeks.

Life cycle – how ladybirds develop

Like all beetles, ladybirds have four stages to their life history: egg, larva, pupa, and adult. Eggs are laid in batches of 20–40. They are often yellowish and elongate. Larvae are grub-like and frequently with spots of yellow, white and orange. As they grow, larvae shed the outer skin four times and after 10–35 days they pupate. Larvae find a safe place (away from other ladybirds), and attach their rear end to a leaf and shed a final larval skin to reveal the pupa. There is no protective cocoon to deter predators and parasitoids (parasitic wasps) but the pupae have an alarm reflex that flicks the body into an upright position when disturbed, startling the attacker. Adults emerge from pupation after two weeks. Initially, the wing cases (elytra) are soft and yellowish but in a few hours they harden and take on adult colouration. Adults must feed to sustain themselves and build up energy for production of eggs or hibernation. Adult ladybirds seek sheltered places to hibernate in autumn and may congregate in large numbers in suitable sites. When warm weather returns in spring ladybirds are eager to feed and find a mate.

Ladybird anatomy – naming the parts

Like most beetles, ladybirds have a pair of hardened wing cases (elytra) that cover and protect both the hind part of the body and wings. The elytra are moveable and open to allow the hindwings to unfold for flight. The head has a pair of short antennae and compound eyes. Between the head and the elytra a large plate, termed the pronotum, protects the neck and first thoracic segment.

Ladybird larvae are generally mobile predators on other insects. Their soft bodies are variously covered with plates on the thoracic segments and with stubby or spiky tubercles on the abdomen. Plates and tubercles may have hairs or spikes rising from them. Larvae are often clearly patterned and readily identifiable to species level.

Adult ladybirds have short legs that can be retracted under the body and fit neatly into grooves on the underside, enabling them to clamp tightly against the substrate.

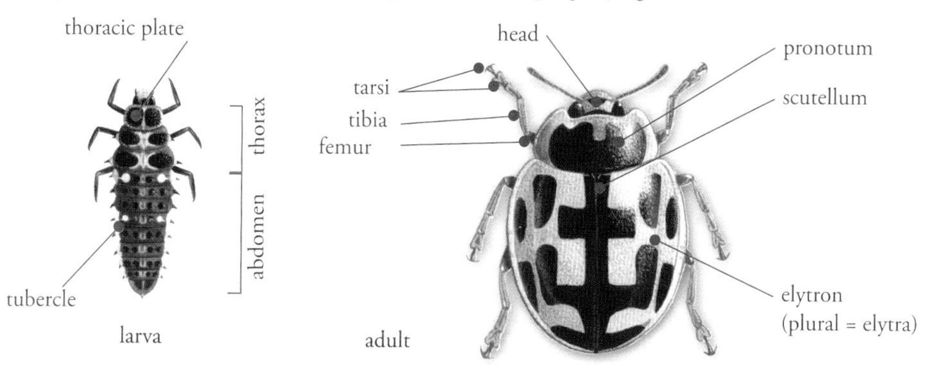

Ladybird food and habitat

Most ladybirds feed on aphids, and in the course of a lifetime an individual can consume hundreds of aphids. Adult ladybirds are relatively long-lived and continue to feed on aphids throughout their lives to sustain energy and to nourish developing eggs. While some ladybirds are associated with a few species of aphid, which in turn are restricted to certain food plants, several species may be found on a wide range of plant species.

Three ladybird species feed on mildews on leaves and stems of plants, and one species feeds on leaves. Adult ladybirds fly readily and may be encountered away their usual habitats but generally occur in specific habitats and they are grouped here accordingly.

Generalists in many habitats – three species

7-spot Ladybird

2-spot Ladybird

Harlequin Ladybird

Habitat: deciduous trees – three species

10-spot Ladybird

Cream-spot Ladybird

Orange Ladybird

Habitat: grasslands – three species

11-spot Ladybird

14-spot Ladybird

22-spot Ladybird

Habitat: conifer trees – four species

Eyed Ladybird

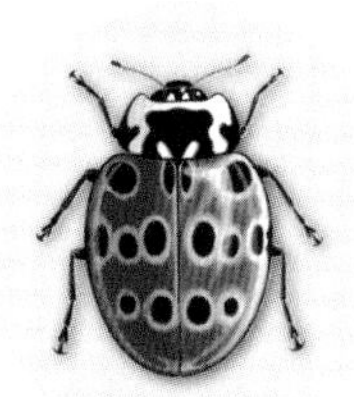

18-spot Ladybird

Larch Ladybird

Habitat: heather – two species

Striped Ladybird

Heather Ladybird

Hieroglyphic Ladybird

Rarely seen – four species

13-spot Ladybird

Water Ladybird

24-spot Ladybird

16-spot Ladybird

Two vagrant species

Two species have been recorded in Ireland that do not nomally occur here. These vagrant species are *Vibidia 12-guttata* (recorded at Barna, 1973) and *Calvia decemguttata* (recorded at Killarney, 1927). Neither is illustrated here.

Larval development

Most ladybird larvae are active and mobile predators of aphids, which makes them a welcome sight in gardens where aphids can be a pest. Larvae begin to feed immediately after hatching; the first meal is the eggshell and sometimes other unhatched eggs in the vicinity. Larvae then seek out aphids. They can tackle prey larger than themselves and may even ride on the back of an aphid as they feed on it. Larvae will eat other ladybird larvae: the non-native Harlequin Ladybird may pose a serious threat to native ladybird populations because it is larger and less selective than native species. Larvae can grow very rapidly given a ready supply of food and favourable temperatures. In some years ladybirds can become very abundant and can form large swarms.

Generalists – three species

7-spot Ladybird

2-spot Ladybird

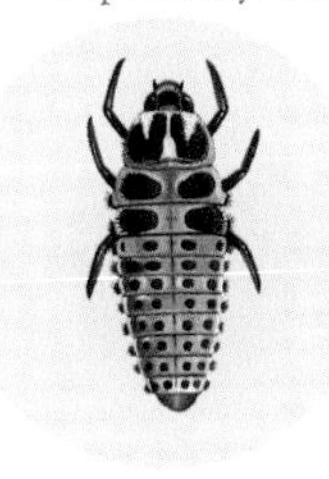

Harlequin Ladybird

Habitat: deciduous trees – three species

10-spot Ladybird

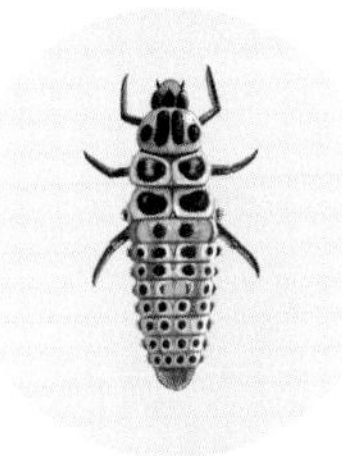

Cream-spot Ladybird

Orange Ladybird

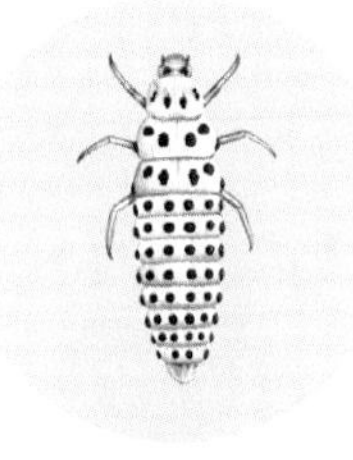

Habitat: grasslands – three species

11-spot Ladybird

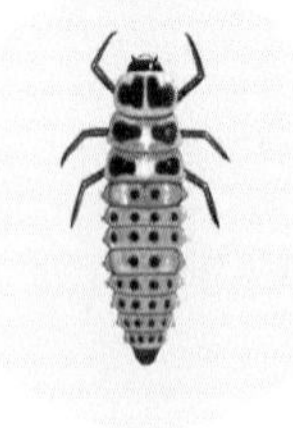

14-spot Ladybird

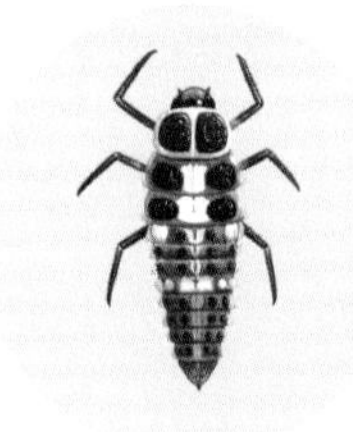

22-spot Ladybird

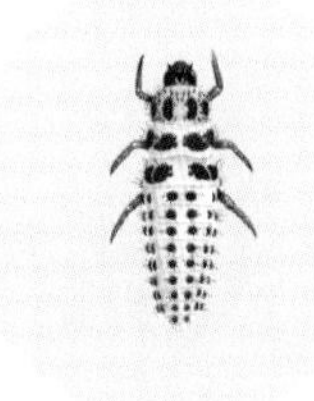

Habitat: conifer trees – four species

Eyed Ladybird

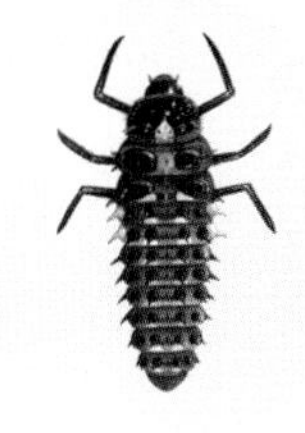

18-spot Ladybird

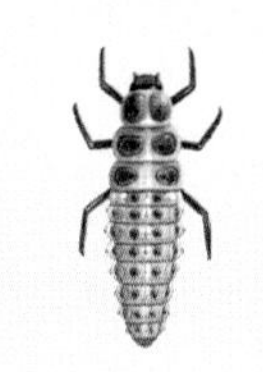

Larch Ladybird

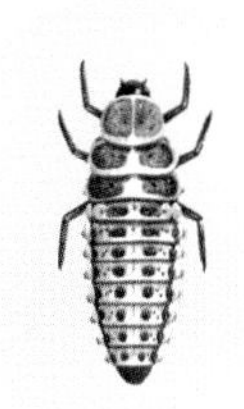

Habitat: heather – two species

Striped Ladybird

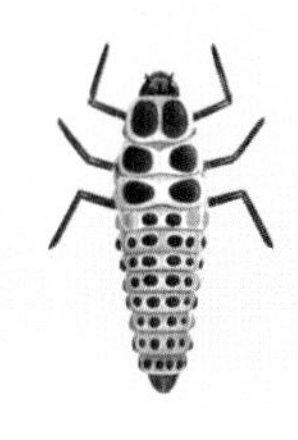

Heather Ladybird

Hieroglyphic Ladybird

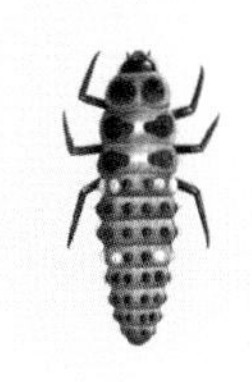

Rarely seen – four species

13-spot Ladybird

Water Ladybird

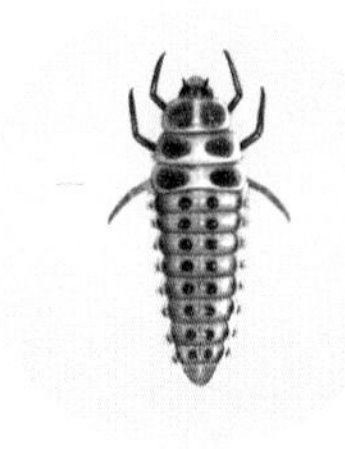

24-spot Ladybird

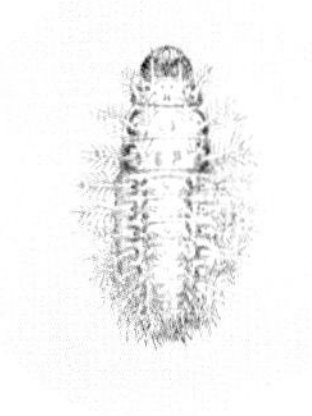

16-spot Ladybird

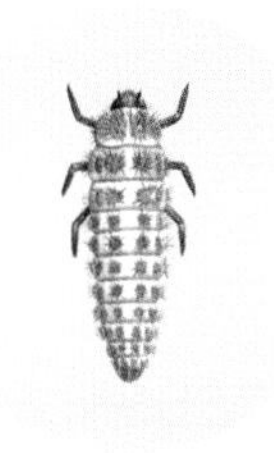

7-spot Ladybird *Coccinella 7-punctata*

DESCRIPTION: *Adult*: Length 5–8mm. This iconic ladybird always has red elytra and seven black spots. Colour variations are rare. *Larva*: Dark grey with two pairs of yellow-orange tubercles on the abdomen. The prothorax is orange with four black plates.
SEASON: Adults overwinter in leaf litter and emerge in spring to feed and breed. Larvae develop during June and July and a new generation of adults matures by late summer. Adults may be found at any time of year but they are most active from spring through to late summer.
NATURE NOTES: Common and widespread in Ireland. Often on herbaceous vegetation. The larvae are usually associated with aphid colonies. Adults and larvae eat a wide variety of aphids on a wide range of plants.
CONFUSION: None.

two pairs of
coloured tubercles
on abdomen

pupa

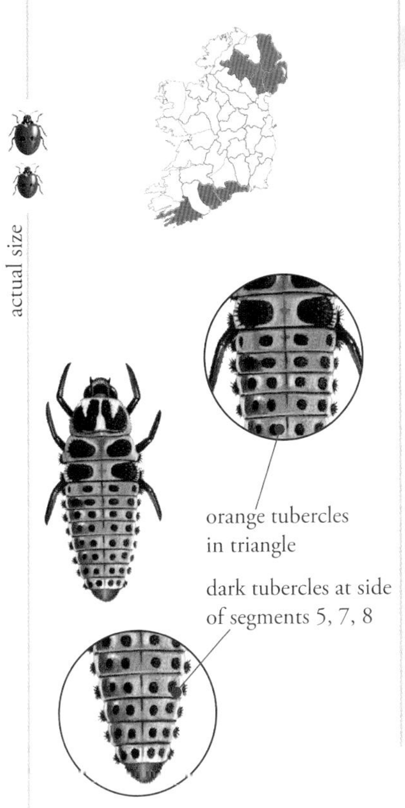

2-spot Ladybird *Adalia 2-punctata*

Description: *Adult*: Length 4–5mm. The typical form has red elytra with two black spots. The pronotum is marked with broad white margins and a dark M-shaped patch. However, colour variations are common and the pronotum and elytra can be black with red spots. The legs are always black. *Larva*: Dark grey with dark tubercles and three orange spots in a triangular formation on the abdomen.

Season: Recorded mainly between May and October but the adults overwinter and may accumulate in numbers on tree trunks and sometimes indoors.

Nature notes: Historically this species was uncommon in Ireland and confined to the south. Since around 1980 it has become established in eastern and northeastern counties. It may be found in a wide range of habitats especially gardens, wetlands, and woods. It feeds on aphids on a variety of plants including deciduous trees, roses, nettles and thistles.

Confusion: The typical form is distinctive but the variable colouration means that individuals could be confused with dark specimens of the 10-spot Ladybird. The 2-spot Ladybird always has black legs and the 10-spot Ladybird has yellow legs.

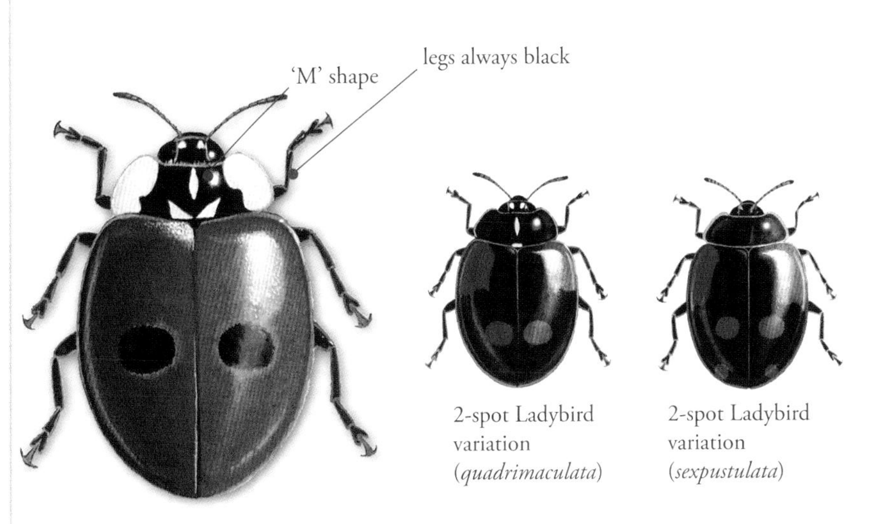

2-spot Ladybird variation (*quadrimaculata*)

2-spot Ladybird variation (*sexpustulata*)

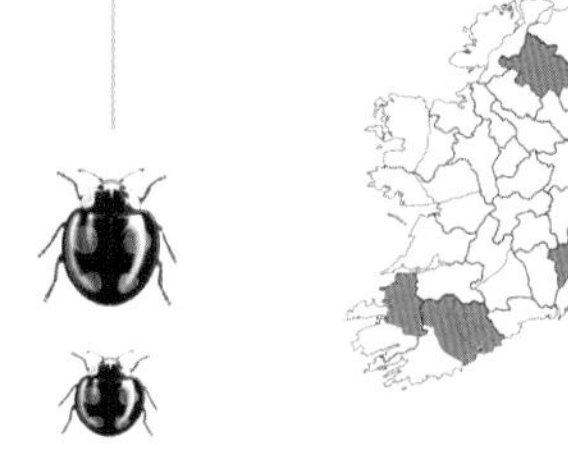

actual size

overall spiky appearance

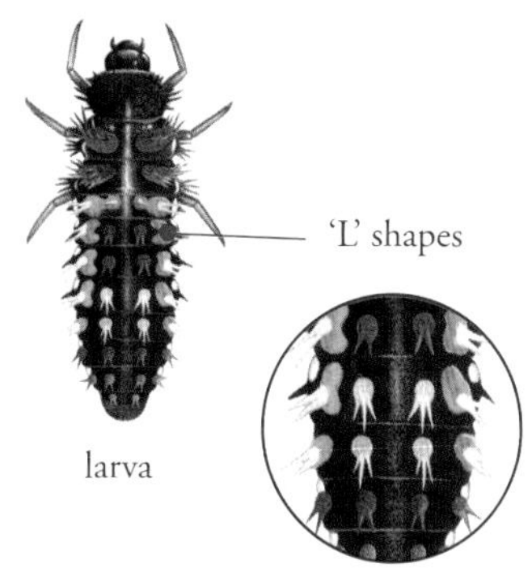

Harlequin Ladybird *Harmonia axyridis*

Description: *Adult*: Length 5–8mm. Colours and markings are very variable. The pronotum and elytra may be yellow, orange, red or black with none to 21 spots. There is a broad ridge running across the tip of both elytra. *Larva*: Dark grey and black with two inverted L-shaped rows of orange tubercles along each side of the body. Overall it is very spiky in appearance.
Season: Adults may be encountered all year round. Larvae develop during the summer months.
Nature notes: This invasive species has recently established breeding populations in Ireland. Regularly reported from Cork but likely to become much more widespread. Native to eastern Asia, it has become widespread in Europe. It occurs in a variety of habitats, including parks, gardens, wetlands, conifer woods and crops. Adults and larvae are predaceous on aphids and other soft-bodied insects, including other ladybirds and aphid predators.
Confusion: Possibly confused with the Eyed Ladybird but this usually has pale rings around the dark spots and is found on conifer trees. The Harlequin Ladybird is larger than most other ladybirds.

Above: the most common forms are those with (left) 2–4 spots on black and (right) 15–21 spots on orange; below: less common variations.

10-spot Ladybird *Adalia 10-punctata*

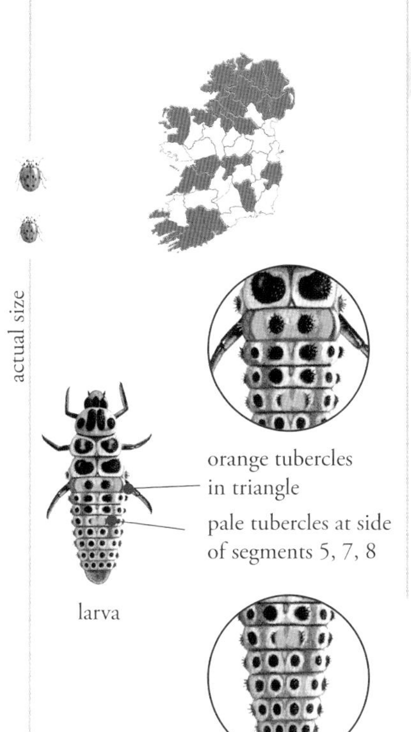

Description: *Adult*: Length 3.5–4.5mm. Typically with ten black spots on orange-red elytra but the colour and markings are very variable. Some forms may be completely black, red, cream or purple, or mainly black with red spots. The pronotum usually has five black marks. The legs are pale or yellow. *Larva*: Grey body with three orange spots, similar to the 2-spot but with pale tubercles on the abdomen.
Season: Adults overwinter in leaf litter and re-emerge in spring. Adults are usually seen between May and October but may be encountered at any time of year.
Nature notes: Widespread and common across Ireland in deciduous trees and shrubs, in woods, hedgerows and herbaceous vegetation. Feeds on aphids on variety of deciduous and coniferous trees.
Confusion: Most easily confused with the 2-spot Ladybird but that species has black legs.

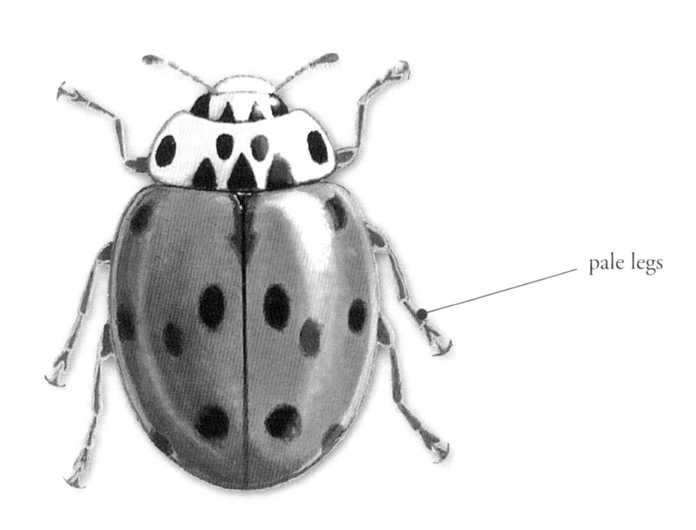

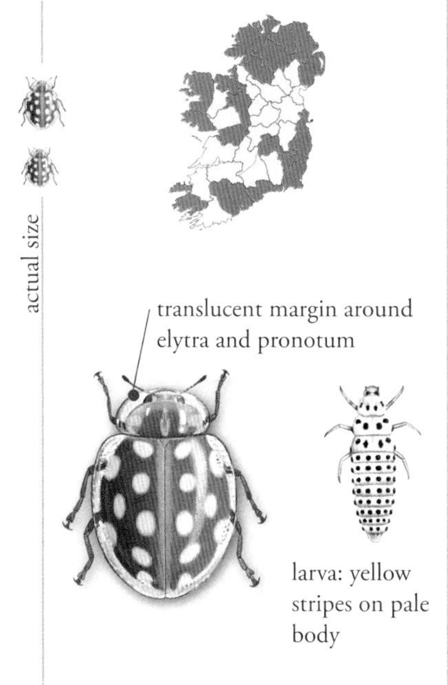

Orange Ladybird *Halyzia 16-guttata*

DESCRIPTION: *Adult*: Length 4.5–6mm. Distinctive because of its orange colour with 12–16 white spots and a translucent margin around the elytra and pronotum. *Larva*: Head, legs and body whitish with bright yellow streaks and strongly marked black tubercles.
SEASON: Adults overwinter leaf litter and emerge later than other ladybirds, around June, and remain active until October.
NATURE NOTES: Widespread and not uncommon on deciduous trees, particularly sycamore and ash. Adults and larvae are thought to feed primarily mildews but they can also eat aphids.
CONFUSION: *Calvia 10-guttata* and *Vibidia 12-guttata* are both orange with white spots and have been recorded in Ireland in the past. Both are very occasional vagrants. Both species are smaller than the Orange Ladybird and have fewer spots.

Cream-spot Ladybird *Calvia 14-guttata*

DESCRIPTION: *Adult*: Length 4–5mm. Almost always maroon-brown with 14 white spots but occasionally variations can occur. The pronotum is brown with pale side margins. A useful distinguishing feature is the row of six spots across both elytra. *Larva*: Dark grey body with black tubercles on thorax and abdomen. A row of pale tubercles along side margins of abdomen and two pairs of pale tubercles on upper surface of first and fourth abdominal segments.
SEASON: Adults overwinter under bark and in leaf litter and are active in late spring and early summer.
NATURE NOTES: Widespread but not common in Ireland. This species lives on deciduous trees and is found mainly in woods and hedgerows and only occasionally on conifers. Adults and larvae feed on aphids and jumping plant lice.
CONFUSION: The 18-spot Ladybird has similar colouration however it is more extensively pale and does not have the row of six spots across the elytra.

11-spot Ladybird *Coccinella 11-punctata*

DESCRIPTION: *Adult*: Length 4–5mm. Elytra are red with 7–11 black spots. Spots may be fused or surrounded by a pale yellow ring, but there is no melanic form. The pronotum is black with white side margins that are often restricted to just front corner. The legs are black. *Larva*: Generally grey body with black plates and tubercles except on first and fourth abdominal segments. First thoracic segment has two large plates.

SEASON: Adults overwinter and are most active between April and October.

NATURE NOTES: Mainly recorded in coastal grassland and estuaries and occasionally inland near rivers and lakes. It usually occurs in plant litter in dry places and it may enter buildings. Both adults and larvae feed on aphids.

CONFUSION: At first glance the adults can look like a small 7-spot but with more spots. Larvae are similar to those of the 7-spot Ladybird but with two plates on the first thoracic segment.

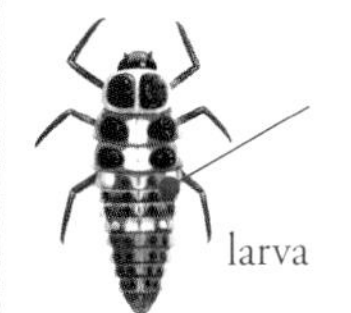

14-spot Ladybird *Propylea 14-punctata*

DESCRIPTION: *Adult*: Length 3.5–4.5mm. The colour of the elytra ranges from pale to dark yellow with black squarish spots. Spots can vary in size and but are always somewhat angular. The pronotum is yellow with a large black mark on the hind margin. *Larva*: Thorax pale with black plates. Abdomen dark grey with white tubercles across the first and fourth segments.

SEASON: Adults overwinter in plant debris and leaf litter and are active between May and October.

NATURE NOTES: Widespread and common in Ireland in a variety of habitats such as trees, shrubs and herbaceous vegetation and especially marshes with willows. Adults and larvae are predaceous on aphids.

CONFUSION: The 22-spot Ladybird is also yellow and black but it is smaller with more numerous and rounded spots.

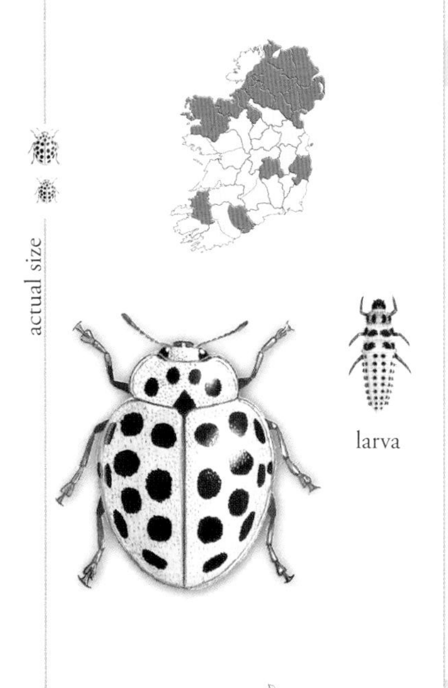

22-spot Ladybird *Psyllobora 22-punctata*

DESCRIPTION: *Adult*: Length 3–4mm. This small ladybird is yellow with 20–22 black spots. There are usually five black spots on the pronotum and colour variations are rare. *Larva*: Distinctive colouration of bright yellow with black spots and black legs.

SEASON: Adults spend the winter sheltering in vegetation and are most frequently seen between April and October but they may be found throughout the year.

NATURE NOTES: Widespread in Ireland, but more common in the east. Often found in grasslands and pastures with umbellifers (especially hogweed), ragwort and thistles, although it does turn up on other plants. Adults and larvae feed on mildews on leaves and stems of herbaceous plants.

CONFUSION: The 14-spot Ladybird is also yellow and black but the spots are larger and somewhat angular.

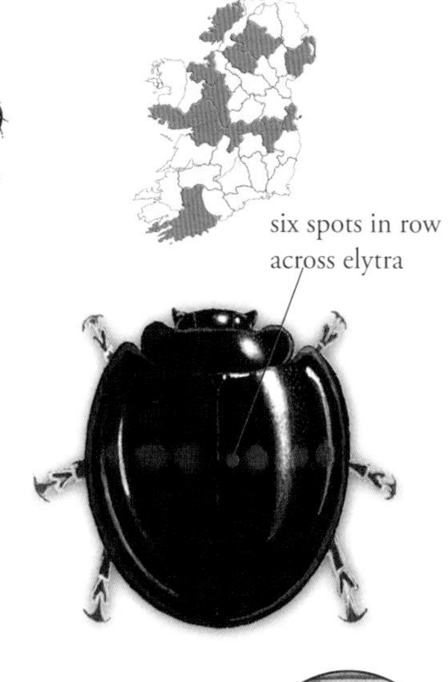

Heather Ladybird *Chilocorus 2-pustulatus*

DESCRIPTION: *Adult*: Length 3–4mm. Distinctly rounded and domed in shape, the elytra are glossy black with 2–6 red spots usually joined in a line. *Larva*: Black and spiky all over except for a pale band across the first abdominal segment.

SEASON: Adults overwinter in leaf litter and evergreen foliage and are active between April and October.

NATURE NOTES: Widespread in Ireland but not common; rare in the north. It is associated with heather, cypress trees and occasionally deciduous trees. Adults and larvae feed on scale insects, aphids and woolly aphids.

CONFUSION: The rounded and domed shape of the Heather Ladybird makes it unlike any other Irish species. The Kidney Spot Ladybird (*Chilocorus renipustulatus*) is similar in shape but has a pair of large round red spots. It is widespread in Britain and may show up in Ireland on deciduous trees. The Pine Ladybird (*Exochomus 4-pustulatus*) is also similar in shape and size but usually has four red spots. It may also show up in Ireland on conifer trees.

very spiky with pale stripe

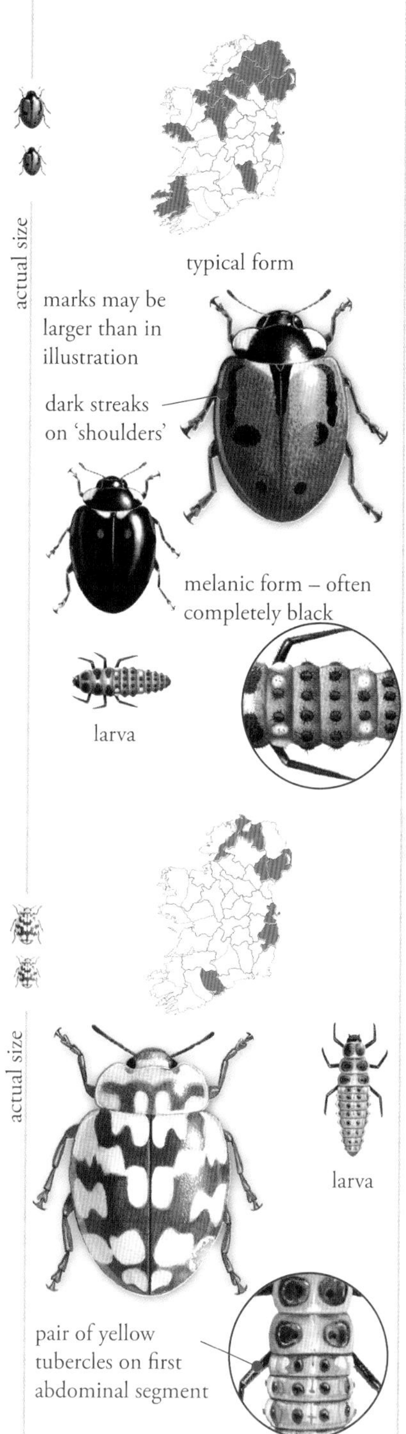

Hieroglyphic Ladybird *Coccinella hieroglyphica*

Description: *Adult*: Length 4–5mm. The typical form is orange-brown with dark streaks on the shoulders and midline of the elytra. The streaks vary in extent and are often fused to form a large black 'M' on the elytra. Almost completely black specimens are frequent in Ireland, with only the front corners of pronotum remaining pale. *Larva*: Head black, body dark grey with black tubercles producing fine black hairs. There are two pairs of pale tubercles on first and fourth abdominal segments and pale patches on second and third thoracic segments.
Season: Adults are most frequently seen in late summer and are most abundant in August.
Nature notes: Widespread but not common in Ireland. It lives in wet places such as margins of raised bogs and lakeshores with mature heather. Adults and larvae feed on the eggs and larvae of the Heather Beetle (*Lochmaea suturalis*) and heather aphids.
Confusion: The melanic form could be confused with melanic 10-spot and 2-spot Ladybirds. Extent and shape of pale markings on pronotum will often help distinguish the species.

18-spot Ladybird *Myrrha 18-punctata*

Description: *Adult*: Length 4–5mm. Medium sized, maroon with 14–18 large cream spots that may be fused. The pronotum is white with a large brown 'M' mark. *Larva*: Pale grey body with black plates and tubercles on the thorax and abdomen. There is a pair of yellow tubercles on the outer part of first abdominal segment.
Season: Adults overwinter in pine trees in crevices in bark and are active during summer.
Nature notes: Uncommon in Ireland and mainly recorded from the east. It lives on mature Scots pine, often high up, but it can occasionally occur on lower branches and sometimes on willow trees. Adults and larvae feed on aphids and woolly aphids in pine trees.
Confusion: The Cream-spot Ladybird has similar colours but with much smaller spots and a row of six spots across the elytra. Larvae of the Larch Ladybird are very closely similar.

actual size

larva

two pairs of orange tubercles on first abdominal segment

Larch Ladybird *Aphidecta obliterata*

Description: *Adult*: Length 4–5mm. Brown elytra with a dark streak on the midline. The pronotum is pale with a dark 'M' shape or variable spots. Melanic forms can occur but have not been seen to date in Ireland. *Larva*: Body is grey with black tubercles except for a pair of bright orange tubercles on each side of the first abdominal segment.
Season: Adult Larch Ladybirds overwinter in crevices in bark on trees and are mainly seen between April and October.
Nature notes: Widespread but uncommon in Ireland. It is associated with conifer trees, most frequently Sitka spruce where it feeds on aphids. Adults and larvae feed on woolly aphids on conifers.
Confusion: Adults are distinctive and not easily confused with other species. Larvae are very similar to 18-spot Ladybird larvae.

actual size

larva

adult: rings absent

Eyed Ladybird *Anatis ocellata*

Description: *Adult*: Length 7.5–8.5mm. Typically burgundy red with 15 black spots surrounded by cream rings. The rings are sometimes absent and occasionally the number of spots may vary from none to 23. The pronotum has a large dark 'M' mark surrounded with white. *Larva*: Dark grey to black and covered in stout black spiky tubercles except for a pair of orange spikes on each side of first two abdominal segments. There is also an orange patch in centre of first thoracic segment.
Season: Adults overwinter in pine needles, soil or moss, underneath host trees. They are active from April to September.
Nature notes: Widespread but quite uncommon in Ireland. Associated with conifers, particularly mature Scots pine and Sitka spruce. Adults and larvae feed on aphids on conifer trees.
Confusion: Possibly confused with the Harlequin Ladybird if the pale rings are absent, but the Harlequin Ladybird has a broad ridge near the tip of each elytron (visible when viewed side on).

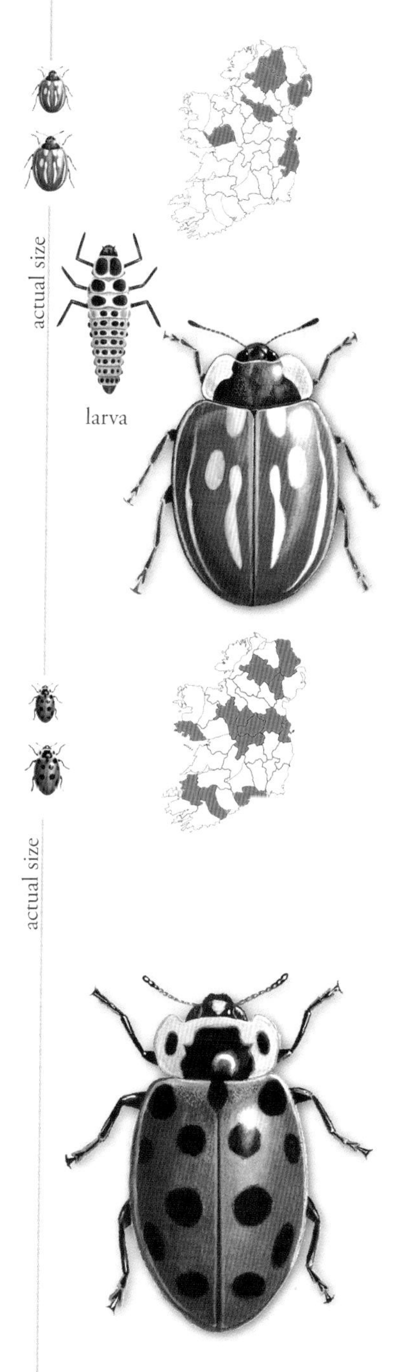

Striped Ladybird *Myzia oblongoguttata*

Description: *Adult*: Length 6–8mm. Reddish to chestnut brown with cream stripes and spots on the elytra. The number of streaks and spots varies but 13 is the norm. The pronotum has broad pale margins and a central dark reddish-brown patch. *Larva*: Grey and generally smooth without spines and with mainly black tubercles. There are orange markings on sides of first, fourth and sixth abdominal segments and on the first thoracic segment just behind the head. The legs are long and black.

Season: Adults overwinter in soil or moss beneath host trees and are active from April to September.

Nature notes: Widespread but uncommon in Ireland mainly on mature Scots pine. Adults and larvae feed on aphids on pines.

Confusion: None, the striped appearance is distinctive.

13-spot Ladybird *Hippodamia 13-punctata*

Description: *Adult*: Length 5–7mm. Orange-red with 13 black spots on the elytra. The pronotum has a large dark central spot and two small lateral spots. The body is rather flattened and elongate. *Larva*: Grey body with black plates and tubercles covered in short stubby hairs. There are pale tubercles on first and fourth abdominal segments.

Season: Adults are mainly recorded in late summer and autumn.

Nature notes: Widespread but not common in Ireland. It lives in wetlands on reeds and grasses. Adults and larvae feed on aphids.

Confusion: Could be confused with the Water Ladybird because of body shape and similar habitat but that species is smaller (4mm) and with 19 spots on elytra and 5 spots on the pronotum.

larva

pale outer pair of tubercles on abdominal segment 1 and row of pale tubercles on segment 4

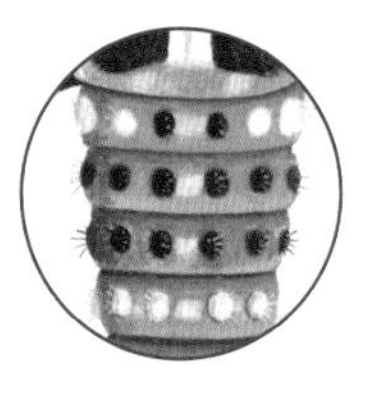

16-spot Ladybird *Tytthaspis 16-punctata*

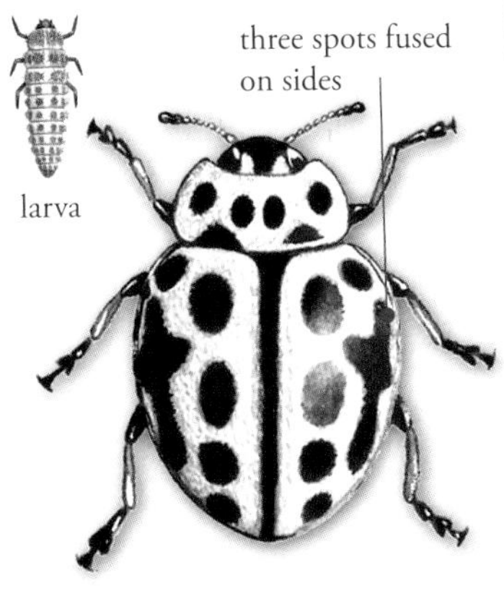

Description: *Adult*: Length 3mm. Creamy-buff with 13–18 black spots. Spots are sometimes joined together, particularly the three along each side. There is also a dark central line on the elytra. The pronotum has four black spots and two dark marks on the hind edge. *Larva*: Pale grey-brown with dark grey tubercles covering in spiky black hairs.
Season: Adults overwinter in low vegetation, plant litter and stone walls and are active from April to September.
Nature notes: Only recorded in Ireland from near Inishtioge, County Kilkenny in 1902. It lives in short vegetation in grassland and meadows. May be overlooked because of its small size. Quite common and widespread in southern Britain. Adults feed on mildews, mites, pollen, nectar and fungi.
Confusion: None.

24-spot Ladybird *Subcoccinella 24-punctata*

Description: *Adult*: Length 3mm. Orange-red with 24 black spots but the spots may be fused and there is a rare melanic form. The pronotum is orange-red with dark spots. Both the pronotum and elytra are covered with short golden hairs. *Larva*: Pale coloured and covered in branched spiny bristles.
Season: Adults spend the winter sheltering in vegetation or grass tussocks and are mainly seen between April and October.
Nature notes: Only recorded once in Ireland, from the Antrim coast in the late 19th century. Lives in grassland and meadows in low growing plants. Adults and larvae eat leaves of a variety of plants.
Confusion: The only hairy ladybird. Its spotted and hairy elytra are unique in Ireland.

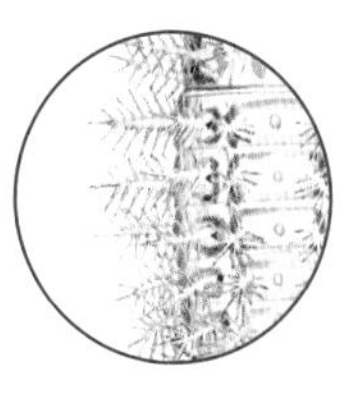

actual size

Water Ladybird *Anisosticta 19-punctata*

DESCRIPTION: *Adult*: Length 4mm. The elytra vary in colour between seasons. In spring and early summer they are red with 15–21 black spots. From late summer to spring they change to pale beige. Body shape is elongated and flat. *Larva*: Thorax pale with dark plates. Abdomen strongly marked with alternating lines of black and white tubercles.
SEASON: Adults are mainly active from May to September and overwinter in reed stems or between leaves. Larvae develop during the summer months.
NATURE NOTES: Only recorded once in Ireland near Kenmare in 1898. It lives on reeds and other waterside vegetation. Widespread and common in England and Wales. Both adults and larvae feed on aphids.
CONFUSION: The 13-spot Ladybird occupies similar habitats but is larger, with fewer spots and no pale form.

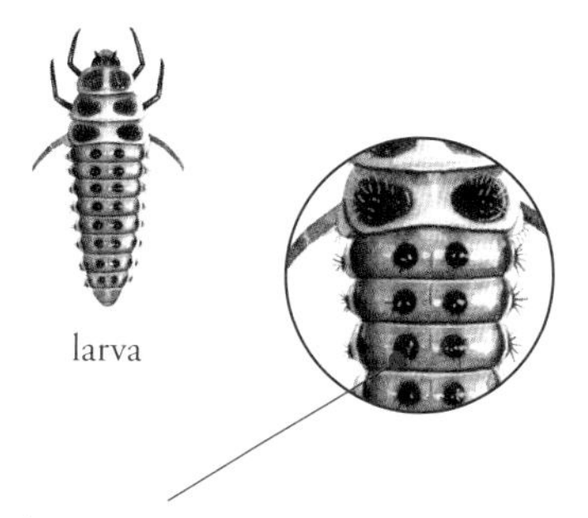

adult: summer colouration

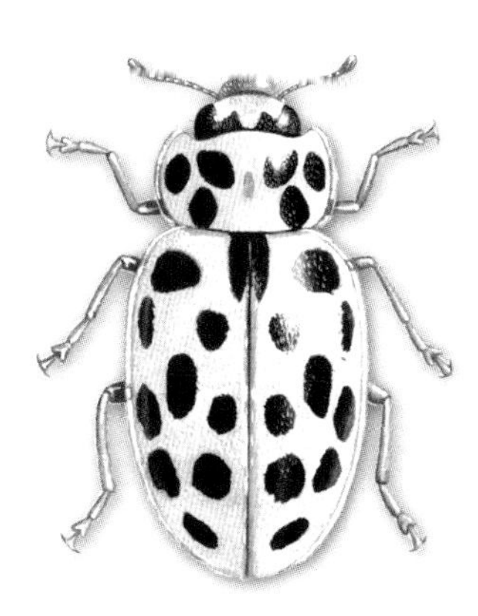

adult: winter colouration

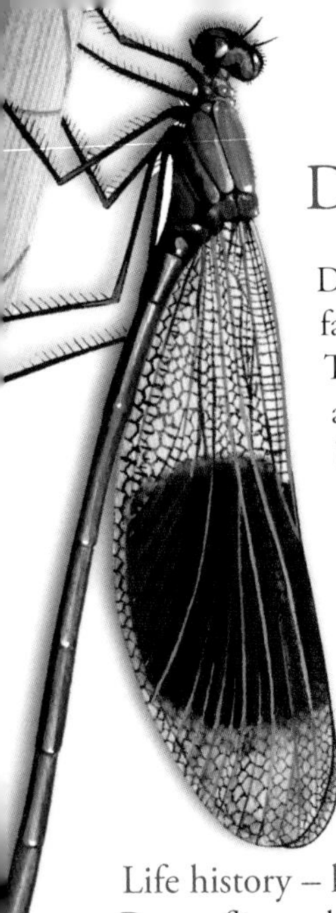

DAMSELFLIES AND DRAGONFLIES

Damselflies and dragonflies are the jewels of the insect world. They are fascinating insects with complex behaviours that are often easy to observe. They belong to the insect order Odonata, meaning 'toothed'. They are amongst the most ancient of all living insects, with fossil remains dating back 300 million years. There are about 6,000 species of dragonfly today but only 24 regularly breed in Ireland. A further eight have been recorded in Ireland on rare occasions.

Damselflies are small, delicate insects, typically up to 3cm long, while dragonflies are larger and more robust. With practice, most species can be identified while flying or at rest. Sometimes close examination is needed, so the insects may be caught and examined before being released. They may be held briefly by holding their wings gently closed between two fingers.

Life history – how dragonflies develop

Dragonflies and damselflies are essentially aquatic insects, spending the majority of their lives as predatory nymphs in fresh or brackish water. The adult phase is brief, lasting only days or weeks at most. Dragonflies have three stages to their life history: egg, nymph and adult. Eggs are either laid individually into plant tissue, soil or moss, or directly into water. Nymphs hatch from the egg and begin to feed. Typical food items are midge larvae, water-lice and small crustaceans. Nymphs are quite sedentary and lie in wait for prey to come within striking distance of their modified mouthparts, sometimes called the 'mask'. This apparatus can be extended at great speed to grasp and stun prey and then retracted to bring prey to the biting jaws. The large, forward-facing eyes enable nymphs to judge prey distance accurately. As they grow, nymphs moult up to 15 times before emerging as adults. Nymph development can take from several weeks in rapidly developing species to several years, depending on the species and environmental conditions. There is no pupa stage and the transition from larva to adult is known as incomplete metamorphosis. The developing wings are clearly visible as 'wing buds' on the thorax. The final nymphal moult takes place out of water and is stimulated by changes in day length and temperature.

Emergence from the water may be synchronised in populations of species such as Emperor Dragonfly, when many individuals emerge over a few weeks in spring. Nymphs haul themselves out of water and may crawl several metres to find a suitable place to perform the delicate act of transformation into a flying insect. The final nymphal skin cracks behind the head and the adult drags itself out of its old skin, now called the exuvia. Newly emerged adult dragonflies are pale at first, with only hints of the final adult patterning. Their wings inflate and become translucent and rigid. Once the body and wings harden, they begin hunting for food whenever fine weather permits.

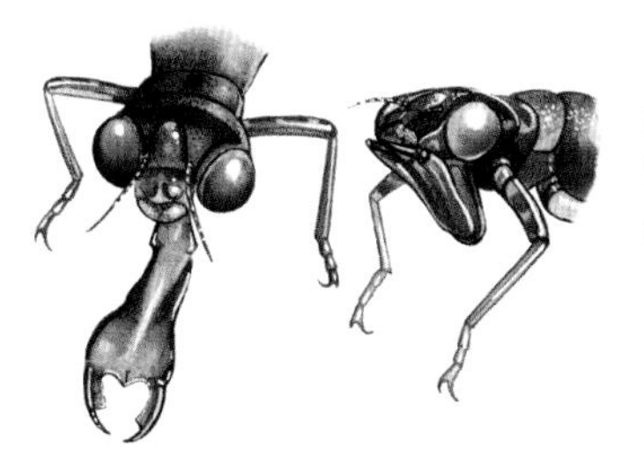

Head of dragonfly showing the 'mask' extended (left) and retracted under the head.

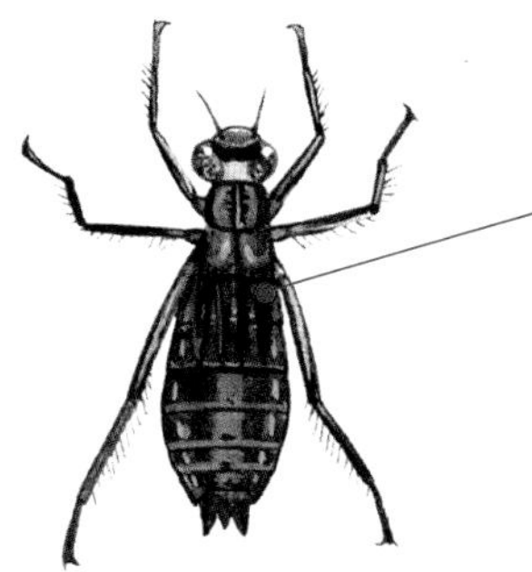

A nymph of a darter or chaser dragonfly showing the squat body shape and long legs.

A damselfly nymph showing the gills extending from the tail - length of insect 30mm.

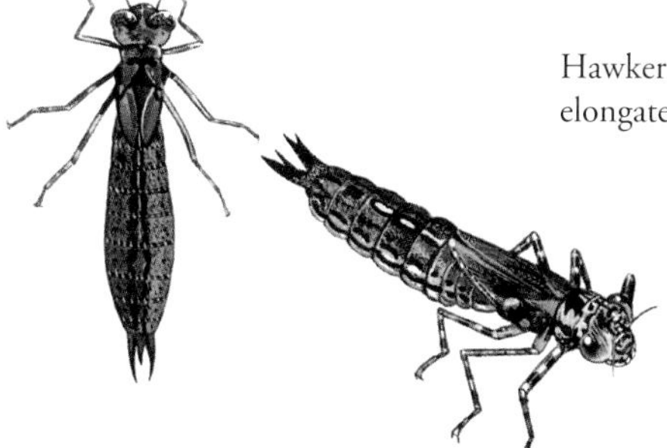

Hawker nymph showing the elongated body shape.

The adult dragonfly becomes sexually mature about a week after emergence from the water. Immature adults lack vivid colouration and usually stay away from water and territorial disputes for several days. Males of many species are territorial and battle with each other to defend a suitable area of breeding habitat. They investigate intruders and any potential mates, whom they will seize or attempt to court. The number of adults found at water is determined by the species' territorial behaviour. Territorial species are always found in smaller numbers than gregarious species, such as the Blue Damselfly, which may be present in hordes.

Mating in dragonflies requires a high degree of physical dexterity and leads to some peculiar behaviour. A male must first transfer sperm from near the tip of the abdomen to accessory genitalia on segments at the front of the abdomen. He then must find himself a mate, which often involves defending an area of breeding habitat from rival males. If a female enters his territory he may court her or simply grasp her by the 'scruff of the neck' with the claspers at the tip of his abdomen. The pair is then said to be 'in tandem'. If the mood takes her, the female curls the tip of her abdomen to meet the male's accessory genitalia and sperm is transferred; this position is known as 'the wheel'. In some species, such as the chasers, the whole mating process takes only a few seconds; at the other extreme, Blue-tailed Damselflies take up to six hours. During mating, the male can remove sperm that the female may have received from previous mates and he may continue to hold her in tandem or guard her as she lays eggs, thus ensuring that his efforts to produce offspring are not in vain.

Adult dragonflies are most active between mid-morning and mid-afternoon, when temperatures are highest. Dragonflies are adept hunters. Their large eyes allow superb visual acuity. In the adult phase the two pairs of large wings, which beat independently, make dragonflies highly manoeuvrable aerial predators.

(L–r): Large Red Damselfly, Ruddy Darter, Black Darter, Blue-tailed Damselfly, Azure Damselfly

Damselfly and dragonfly anatomy

The head has very large compound eyes and short antennae. The thorax is large and houses the huge flight muscles, which power the wings. The leading edge of the wing is called the costa and there is a dark spot near the tip called the pterostigma. There are often prominent stripes on the upper surface of the thorax called antehumeral stripes. The legs are long and slender and often fringed with short bristles; they are held tucked under the body in flight and used to grasp prey. The abdomen is long and slender and comprised of ten visible segments and appendages on the tail end. Males have claspers for grasping females just behind the head. Females have an egg-laying organ that may be used to insert eggs into floating or submerged vegetation.

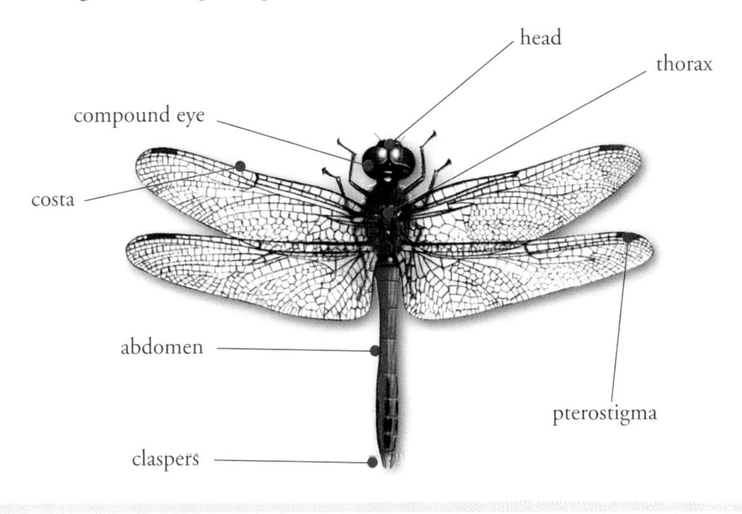

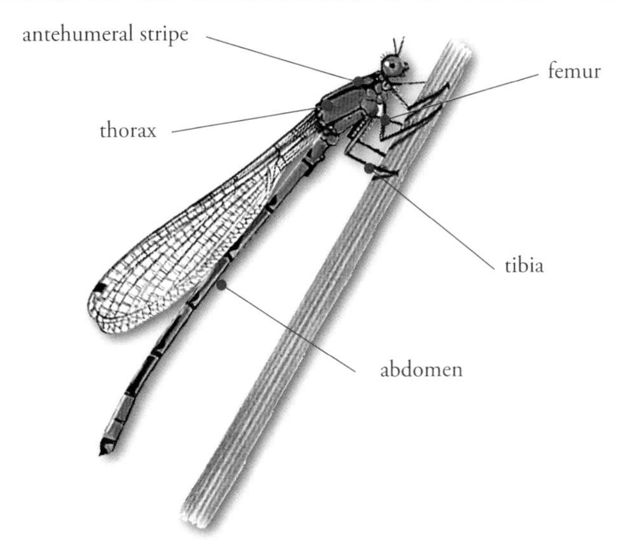

Identifying Irish damselflies

There are eleven species of damselfly in Ireland: six blue species, four metallic-green species and one red species. When identifying damselflies, note that females and immature males are very variable and more difficult to identify. Both males and females are likely be present at breeding sites. Another useful identification aid for Irish damselflies is habitat association. Here are a few lists of Irish damselflies by breeding habitat:

Typical habitat associations

Small lakes and ponds: Emerald Damselfly, Scarce Emerald Damselfly, Large Red Damselfly, Common Blue Damselfly, Azure Damselfly, Variable Damselfly, Irish Damselfly

Bog/fen: Large Red Damselfly, Scarce Blue-tailed Damselfly, Variable Damselfly, Irish Damselfly, Azure Damselfly, Blue-tailed Damselfly

Streams and rivers: Banded and Beautiful Demoiselles, Large Red Damselfly

Large lakes: Common Blue Damselfly, Blue-tailed Damselfly

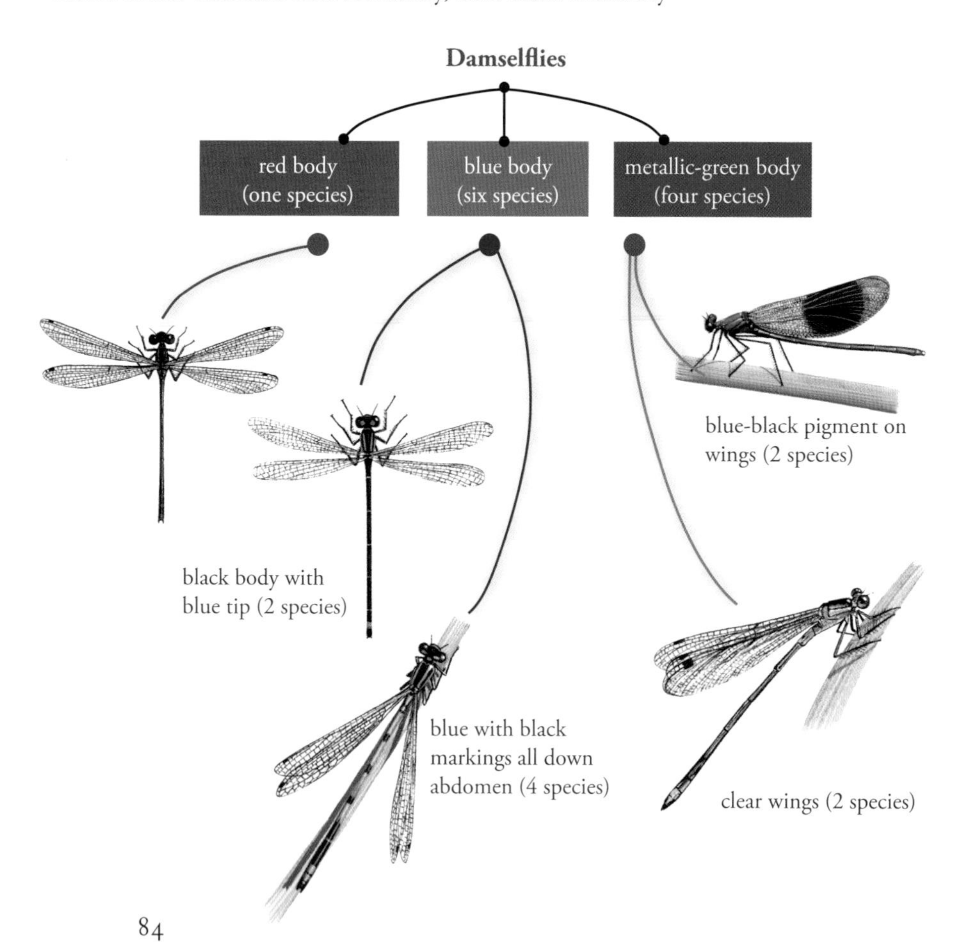

Large Red Damselfly
Pyrrhosoma nymphula Earr-rua an Earraigh

Description: *Male*: Length 36mm. Eyes, thorax and abdomen red above with black markings on the thorax and tail. Underside of head and thorax yellowish. Antehumeral stripes are yellow when immature. *Female*: There are three colour forms and most have yellow antehumeral stripes and are darker on the abdomen than males.

Season: Emergence can be as early as the end of April and most emerge within a three-week period in response to rising temperatures. Peak abundance is in June and by mid-August few are seen.

Nature notes: Widespread and common throughout Ireland. Prefers acidic water and typical habitats include shallow bog pools, small lakes and small streams. Males defend a territory from a perch and investigate passing damselflies. Mating takes about 15 minutes and eggs are laid, in tandem, onto submerged plants or the underside of floating leaves. Nymphs develop over two summers. At the end of the second summer they enter a resting phase ready for emergence in spring.

Confusion: There is no other red damselfly in Ireland and this species is usually the first dragonfly to emerge in late spring.

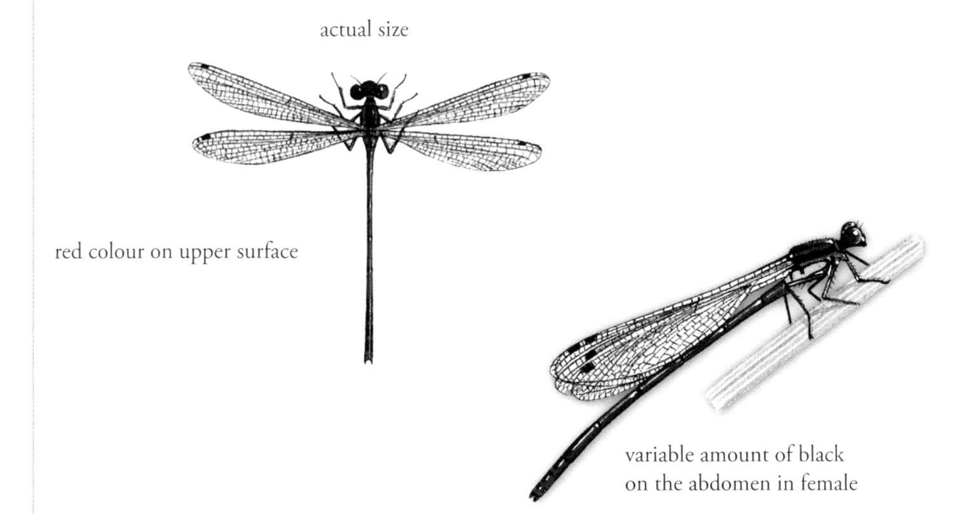

Blue-tailed Damselfly
Ischnura elegans Rinnghorm Coiteann

DESCRIPTION: *Male*: Length 31mm. Mature male is blue and black on the thorax with blue antehumeral stripes. The abdomen is yellowish beneath and black above with the blue tail only on segment 8. Immature males are initially green on the thorax but turn blue over about 10 days. *Female*: Similar to male. The tail may be reddish, although it is occasionally all black. There are three colour variants and two immature colour forms. The thorax may be reddish-pink or violet and may sometimes lack antehumeral stripes.
SEASON: Adults emerge from early May through to the end of August with peak numbers between June and early August.
NATURE NOTES: Widespread and common in most freshwater habitats in Ireland although it is less abundant near flowing water or exposed sites. It is most abundant in sheltered lowland ponds and lakes. Males are territorial and mating can take several hours. Females lay eggs alone, inserting them into stems of emergent plants just below the water surface. Nymphs develop over one or two years in shallow waters.
CONFUSION: The Scarce Blue-tailed Damselfly is similar but with the blue tail on part of segment 8 and all of segment 9.

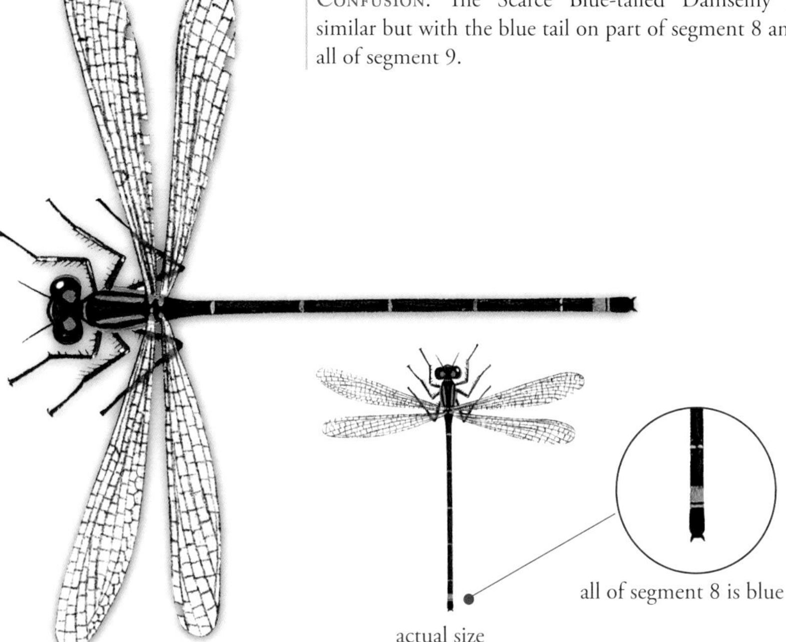

Scarce Blue-tailed Damselfly
Ischnura pumilio Rinnghorm Beag

Description: *Male*: Length 26–31mm. Mature male is mainly black on the dorsal surface of abdomen with the blue tail only on part of segment 8 and all of segment 9. Thorax sides are initially pale but turn green then blue over a few days. *Female*: Abdomen is completely black above and lacks the blue tail. Immature females are marked with orange on the thorax and underside of abdomen but as they mature they become greenish.
Season: Emergence begins in May with peak around mid-June and the flight season lasts until early September.
Nature notes: Widespread but very localised in shallow seepages and the margins of large limestone lakes, quarries and springs. Territorial males perch and pursue passing females. Mating takes place on vegetation and females then lay eggs unaccompanied, inserting eggs into stems of submerged plants. Nymphs develop rapidly in warm shallow water.
Confusion: Blue-tailed Damselfly is similar but with the blue tail on segment 8.

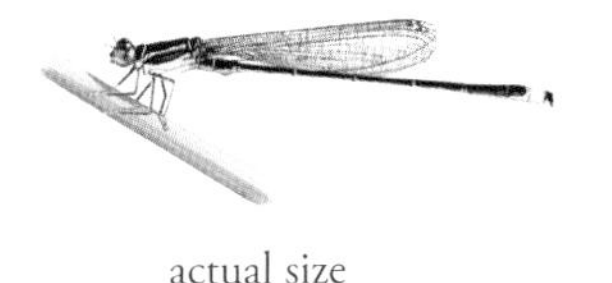

actual size

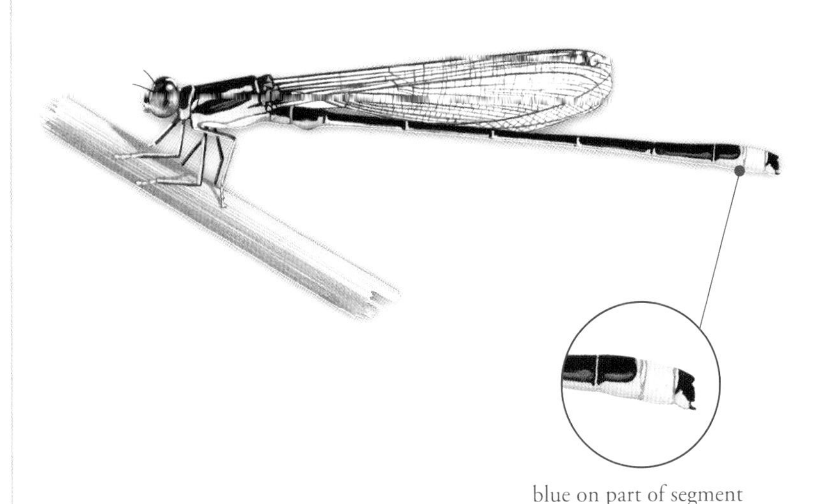

blue on part of segment 8 and all segment 9

Common Blue Damselfly
Enallagama cyathigerum
Goirmín Droimriabhach

Description: *Male*: Length 32mm. Mature males are sky blue and black. Thorax has distinct broad antehumeral stripes and lacks a black 'spur' on the side. The second abdominal segment has a distinctive club-shaped mark. *Female*: Mature females are blue, brown or green with black markings. Abdominal segments have spear-shaped black markings.
Season: Emergence begins as early as April. Peak abundance is in June and July but the flight season is long and can extend to the end of September.
Nature notes: Widespread and common in Ireland. Most abundant around lakes, but also breeding in ponds, quarries, slow rivers and canals. Females feed away from water. Males congregate around water margins. Tandem pairs are frequent at breeding sites and egg laying usually takes place in tandem. Females insert eggs into stems of submerged plants and can submerge. Nymphs are ambush predators and development usually takes one year but, in unfavourable conditions, may take longer.
Confusion: Variable, Azure, and Irish Damselflies could be confused but they have a spur on the thorax, narrower antehumeral stripes and lack the club shape on the second abdominal segment.

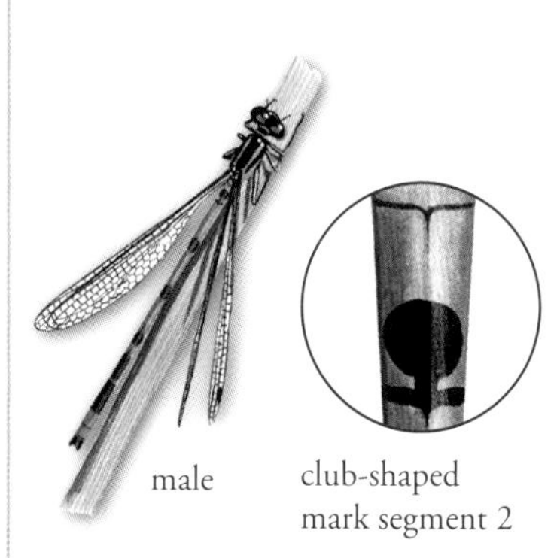

male

club-shaped mark segment 2

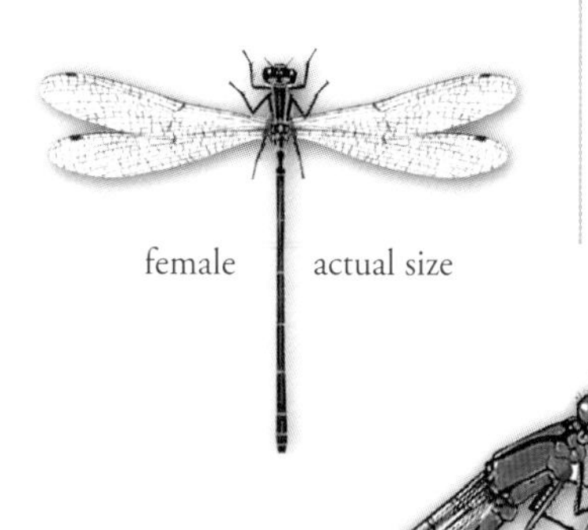

female

actual size

thick antehumeral stripes

all-blue upper surface to segments 8 and 9

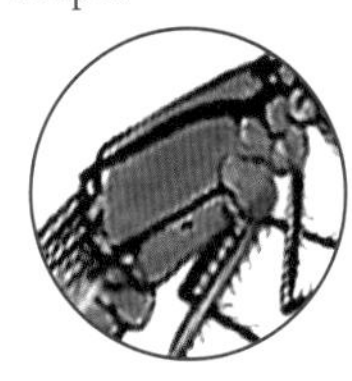

no black spur

Azure Damselfly
Coenagrion puella Goirmín Spéiriúil

Description: *Male*: Length 33mm. Mature individuals are blue and black with complete antehumeral stripes and a distinctive U-shaped mark on the second abdominal segment. *Female*: Stouter than males and mainly black on the dorsal surface of abdomen. Most females are green or (less commonly) blue on the abdomen and thorax.

Season: Adult emergence begins in early May and the flight season lasts until early September.

Nature notes: Widespread and common across much of Ireland, breeding in sheltered lowland lakes and ponds with emergent vegetation. Males perch and patrol the waters edge in search of females. Mating takes about 30 minutes followed by egg laying in tandem into submerged plants. Nymphs develop over one or two years.

Confusion: Females are easily confused with females of other damselfly species. Males, however, are more easily distinguished. The Common Blue Damselfly lacks the black 'spur' on the thorax. The Variable and Irish Damselflies have differently shaped markings on the second abdominal segment.

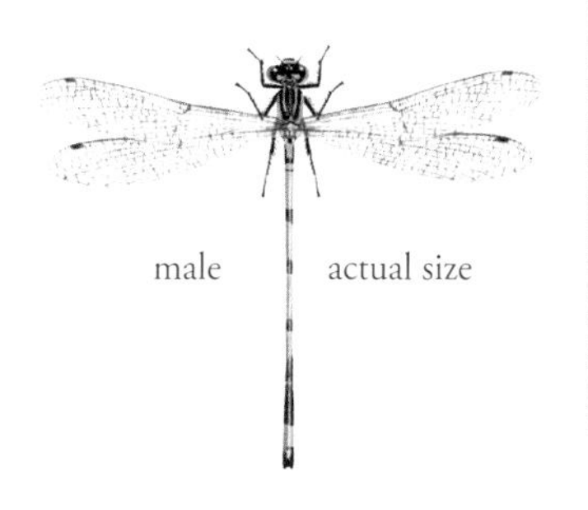

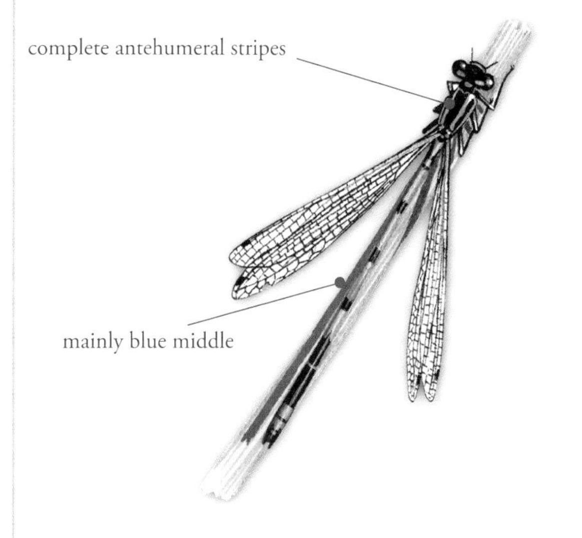

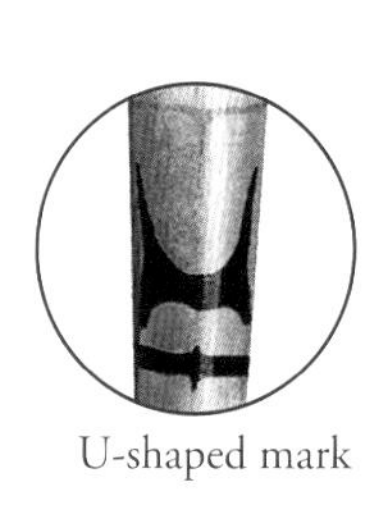

U-shaped mark

Variable Damselfly
Coenagrion pulchellum Goirmín Luainach

DESCRIPTION: *Male*: Length 33mm. Mature males are blue and black with a wine-glass-shaped mark on the second abdominal segment. Antehumeral stripes are sometimes interrupted. *Female*: The blue form has a blue or green thorax and a blue-and-black abdomen. The dark form is almost entirely black on the upper surface of the abdomen.
SEASON: Emergence begins in late April with peak numbers between June and late July with few surviving into August.
NATURE NOTES: Widespread and common in Ireland, occurring in most freshwater habitats including ponds, ditches and slow flowing water. Females feed and shelter away from water and only approach it to breed. Males are territorial and perch on waterside vegetation. After mating, the couple fly in tandem seeking suitable egg-laying sites, often on the underside of dead or decaying leaves. Nymphs take one or two years to develop.
CONFUSION: The Common Blue, Azure and Irish Damselflies are all similar. Male Variable Damselflies are best distinguished by the combination of the black 'spur' on the thorax and the wine-glass-shaped mark on the second abdominal segment. Females are most easily identified by association with males.

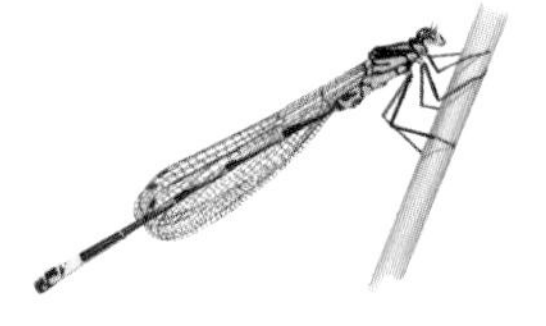

actual size

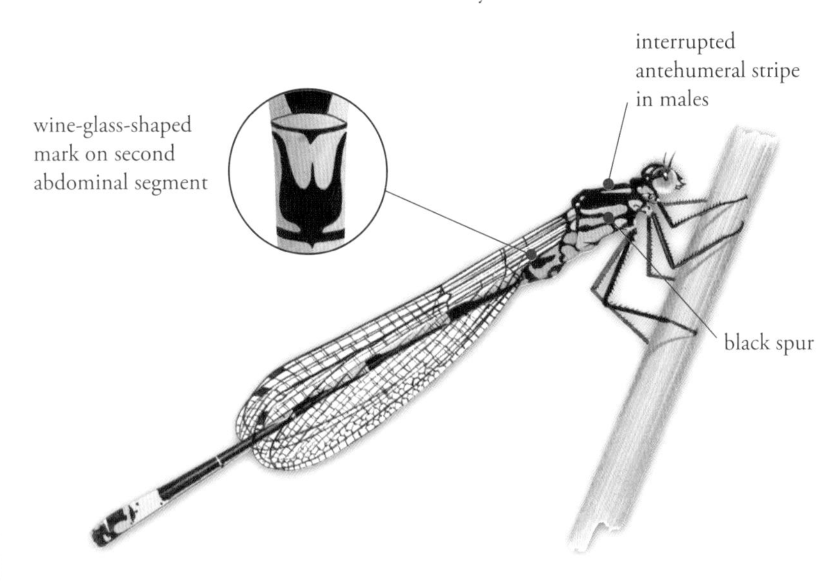

Irish Damselfly
Coenagrion lunulatum
Goirmín Corránach

DESCRIPTION: *Male*: Length 31mm. Mature males are blue and black above with segments 8 and 9 all blue above. The underside of the body is green. Second abdominal segment markings are a black crescent near the hind margin and a black bar along each side. *Female*: Mature females are mainly dark on the upper surface, yellow underneath and with yellow antehumeral stripes and thorax sides.

SEASON: Emergence begins in mid-May with peak numbers in late June with few records during August.

NATURE NOTES: First recorded in Ireland 1981 in Sligo. Previously overlooked, it is widespread but uncommon in the northern half of Ireland as far south as the Burren. The main habitat is sheltered lakes or large pools on cutover bogs. Males perch on floating leaves and congregate at breeding sites. Egg laying occurs in tandem; females climb down stalks of pondweeds and insert eggs into stems. Nymphs take one year to develop.

CONFUSION: The Common Blue, Azure and Variable Damselflies are all similar. The black 'spur' on the thorax and crescent-shaped mark on the second abdominal segment are distinctive. Females are difficult to tell from other females and are most easily identified by association with males.

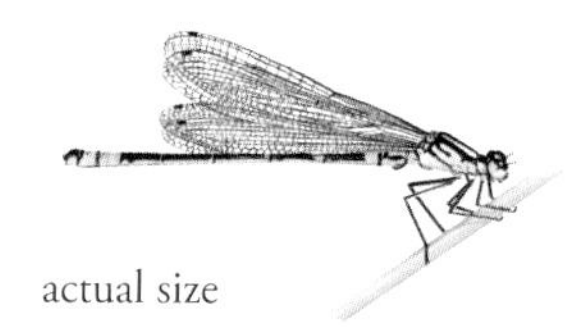

actual size

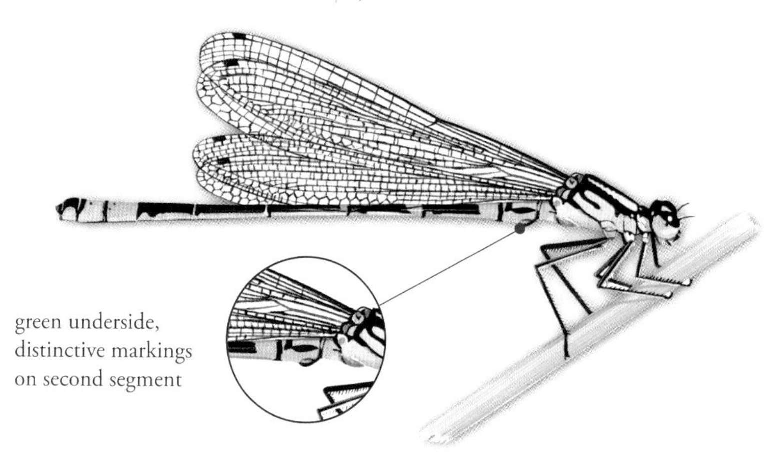

green underside, distinctive markings on second segment

Banded Demoiselle
Calopteryx splendens Brídeog Bhandach

DESCRIPTION: *Male*: Length 45mm. The body is dark metallic blue or green. Wings have a dark blue-black band and unpigmented areas at the base and tip. *Female*: The body is a deep metallic green and the wings are tinted green. There is a small pale stripe on the upper surface near the tail.
SEASON: Early individuals appear at the end of May. Flight season is long and extends to the end of September.
NATURE NOTES: Widespread throughout Ireland in slow-flowing rivers and canals and sometimes lakes. It is common in large, slow rivers, particularly with emergent vegetation. Males defend a stretch of river and attract females by flicking their wings in front of them. Females lay eggs into plant tissues. Nymphs live on the river bottom and burrow into sediment in winter.
CONFUSION: Beautiful Demoiselle males have all, or almost all, of the wings dark. Females have brown-tinted wings and brown stripe near the tail.

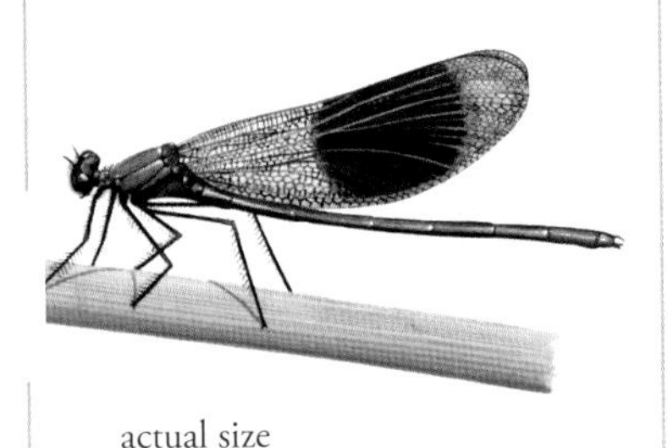

actual size

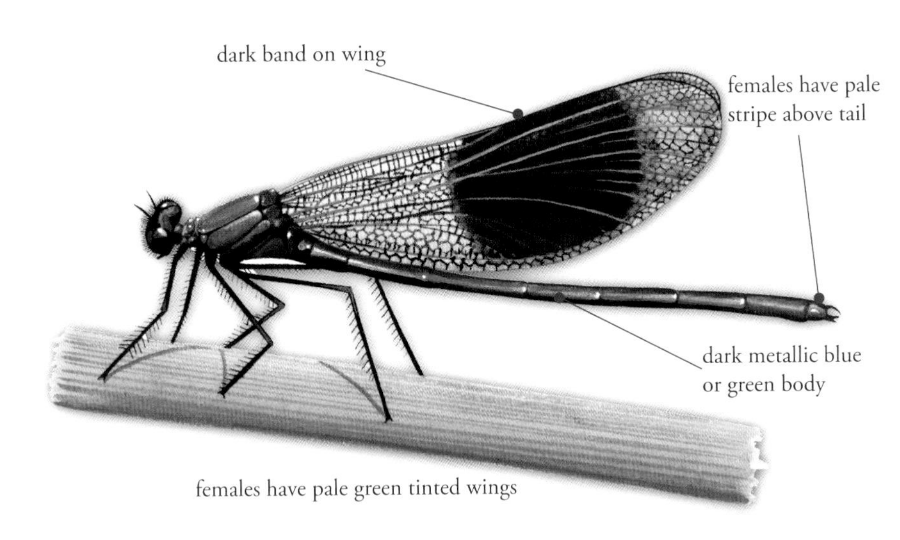

actual size

Beautiful Demoiselle
Calopteryx virgo Brídeog

DESCRIPTION: *Male*: Length 45mm. Mature male has dark brown/black-tinted wings with blue iridescent veins. The body is dark metallic blue and the black legs are long. Newly emerged specimens have pale wings. *Female*: The body is metallic green and the wings are tinted brown. There is a brown stripe on the upper surface near the tail.

SEASON: Emergence begins in May with peak numbers between June and mid-August. Flight season lasts until early September.

NATURE NOTES: Widespread across the southern half of Ireland. It breeds in rivers and streams with moderate to fast flow with gravel or stony beds. Adult males defend patches of emergent vegetation and perform a fluttering flight to impress females. Males guard females during egg laying, which takes place into submerged vegetation such as bur-reed. Nymphs are distinctively long and narrow with long antennae and legs. They burrow into substrate over winter and take two years to develop.

CONFUSION: Male Banded Demoiselles have wings partly dark, females have greenish rather than brown tinted wings and pale rather than brown stripe on the upper surface near the tail.

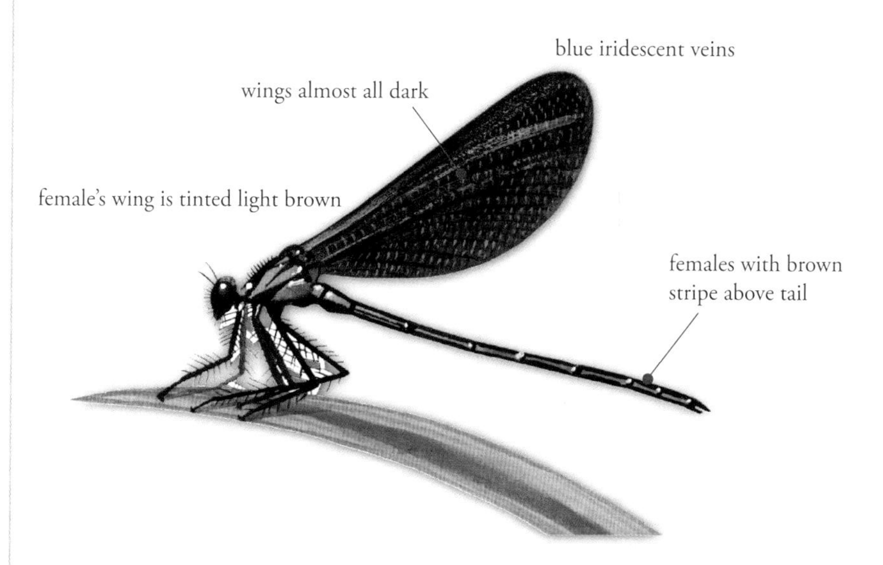

Emerald Damselfly

Lestes sponsa Spré-eiteach Coiteann

Description: *Male*: Length 38mm. The body is metallic green with powder-blue on first two and last two segments of abdomen which develops after few days. The wings are clear and held partly open when at rest. *Female*: Stouter body than male and duller green. No powder blue on the body. Rounded, teardrop-shaped marks on abdominal segment 2.

Season: Early individuals emerge around mid-June with peak numbers in July. Flight season lasts until early September.

Nature notes: Widespread and common in acidic ponds and small lakes or vegetated margins of larger lakes. It flies only short distances and is a relatively sedentary species. Males are territorial and congregate at breeding sites. Tandem pairs are frequent and eggs are laid in rows in stems of emergent vegetation. Adults may submerge entirely while laying eggs progressively down a stem. Nymphs are active predators and move about amongst vegetation, feeding on small crustaceans and midge larvae.

Confusion: Males of the much rarer Scarce Emerald Damselfly have only half of the second abdominal segment covered in powder blue. Female Scarce Emerald Damselflies have square-shaped marks on segment 2.

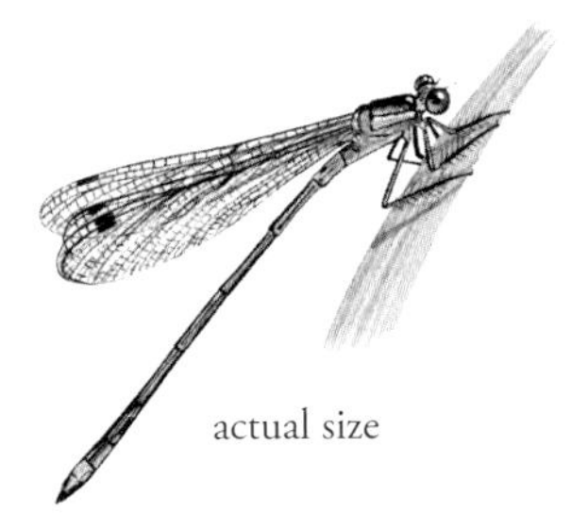

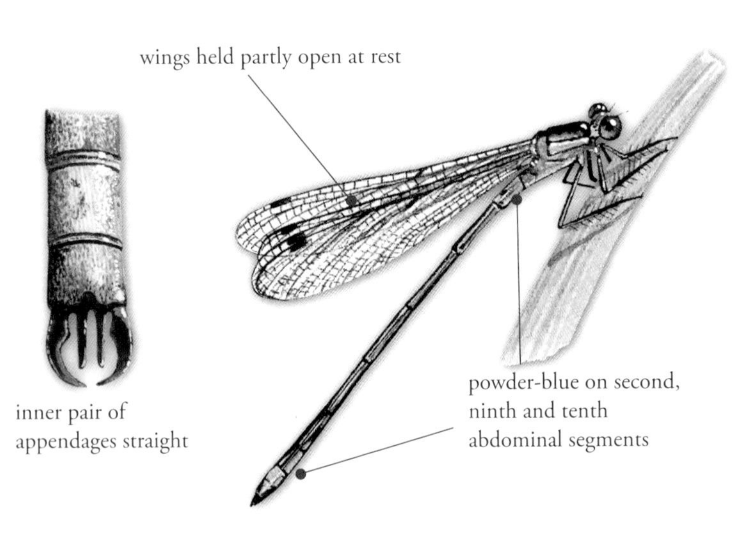

Scarce Emerald Damselfly
Lestes dryas Spré-eiteach Turlaigh

DESCRIPTION: *Male*: Length 34–37mm. The body is metallic green with clear wings held partly open at rest. Mature male has bright blue eyes and develops a powder-blue colouration on half of the second abdominal segment as well as the ninth. *Female*: More robust than male, eyes green, metallic-green body lacking powder blue of males. Square spots on segment 2 of abdomen.

SEASON: Emergence begins around mid-June. The flight season lasts into early September.

NATURE NOTES: Very localised in Ireland in temporary or fluctuating bodies of water including turloughs, disused canals and small ponds, mainly in western and central counties. Mating takes up to two hours and eggs are laid, usually in tandem, into stems, sometimes above water level. Eggs remain dormant until the following spring. Nymphs develop rapidly in temporary water, maturing as water levels drop.

CONFUSION: Possibly confused with the Emerald Damselfly but in males the second abdominal segment is entirely powder blue. Female Emerald Damselflies have round spots on the second abdominal segment.

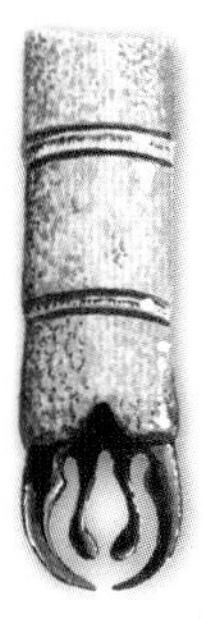

inner pair of appendages curve inwards at tip

Male with power-blue on only half of second segment

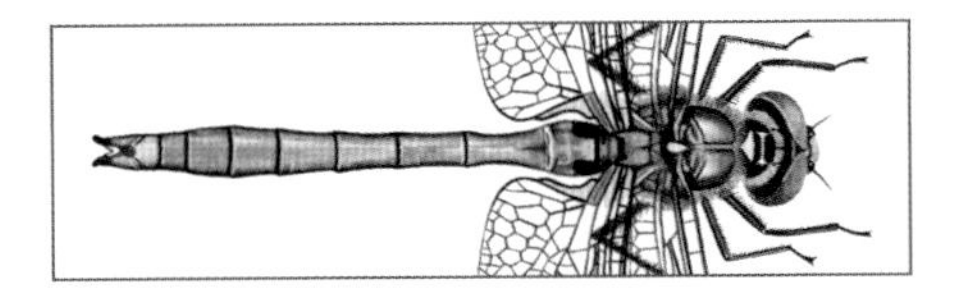

Downy Emerald (48mm)
Key identification feature: club-shaped abdomen and slightly curved appendages.
Flight period: May–July.

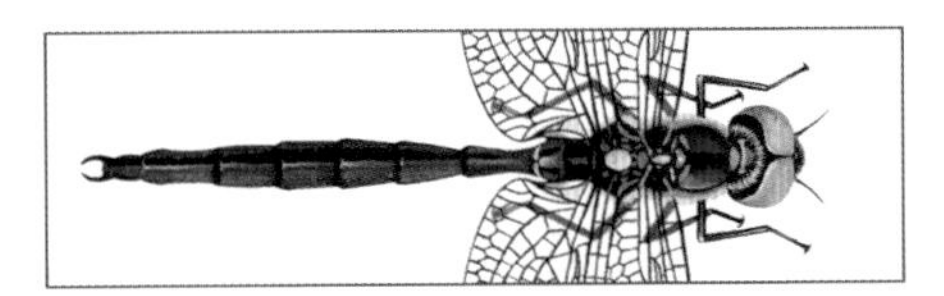

Northern Emerald (50mm)
Key identification feature: males have a waisted abdomen and calliper-shaped appendages.
Flight period: June – July.

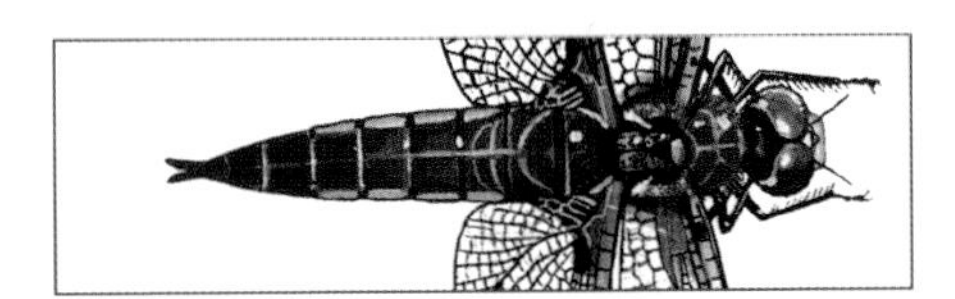

Four-spotted Chaser (39–48mm)
Key identification feature: spots on wing margins.
Flight period: May–August.

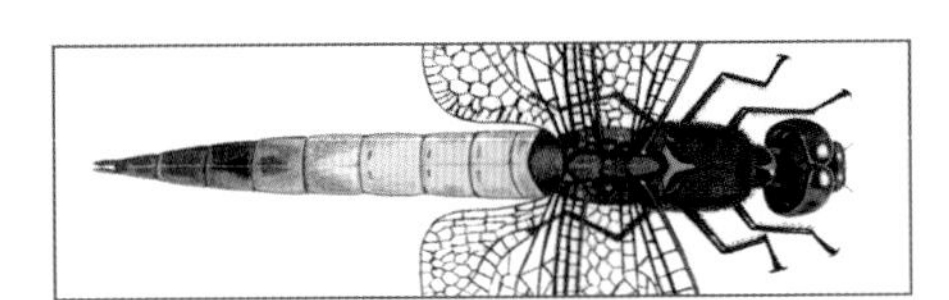

Black-tailed Skimmer (44–49mm)
Key identification feature: black tip on blue abdomen.
Flight period: May – August.

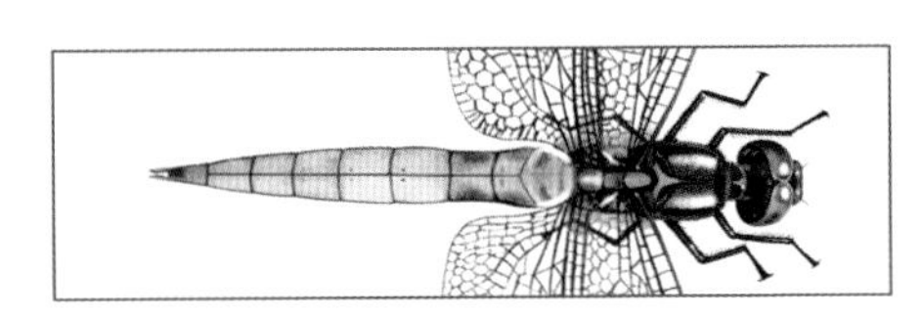

Keeled Skimmer (40–44mm)
Key identification feature: blue abdomen without black tip.
Flight period: June – August.

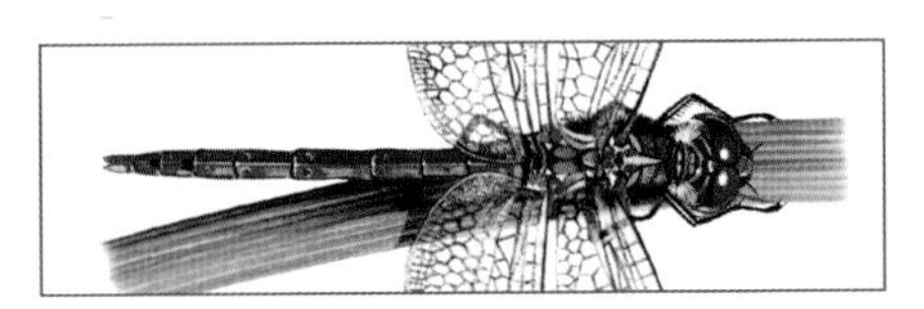

Common Darter (40mm)
Key identification feature: yellow stripe on black legs.
Flight period: June–October.

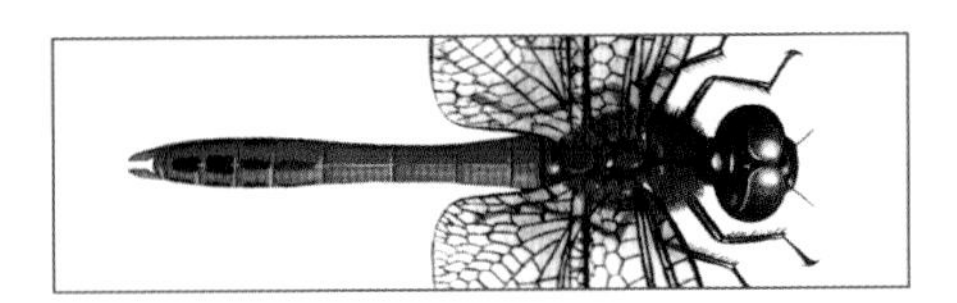

Ruddy Darter (34mm)
Key identification feature: waisted abdomen and all-black legs (no stripes).
Flight period: June–September.

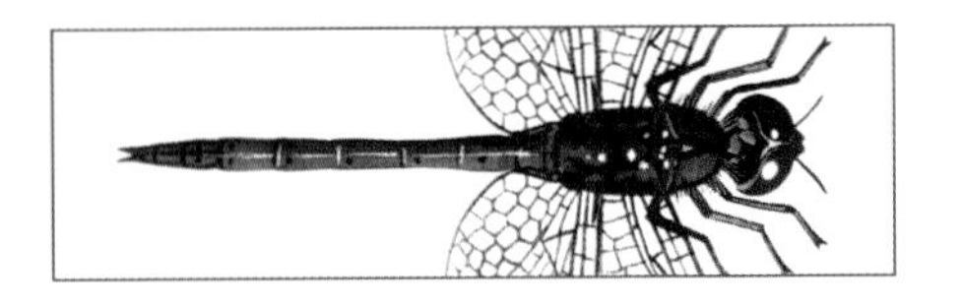

Black Darter (32mm)
Key identification feature: males all black.
Flight period: July–October.

Identifying Irish dragonflies

Identification of Irish dragonflies relies on size, colour and behaviour. Habitat and flight season are important for some species but the larger dragonflies are highly mobile and can roam far from their breeding habitat. Note that the colouring of dragonflies changes as the insect ages. The illustrations show the mature adult males only. Males and females are often different in appearance so check the accompanying text for female identification. The maps show the vice-counties with records for each species.

Hawkers and Emperor Dragonflies

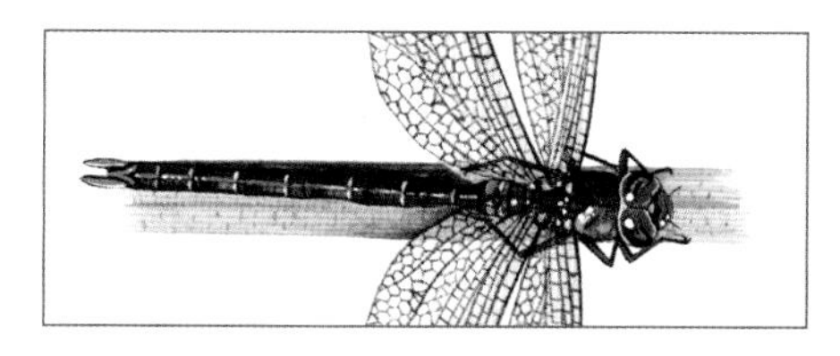

Brown Hawker (73mm)
Key identification feature: wings tinted amber.
Flight period: mid-June – late September

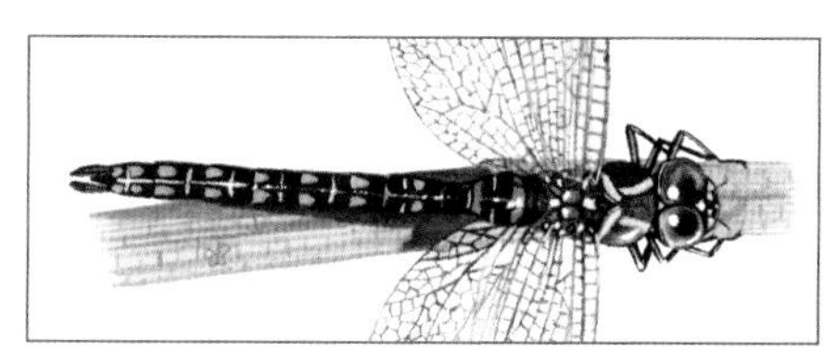

Common Hawker (74mm)
Key identification feature: yellow costa (vein running along wing edge).
Flight period: June–October.

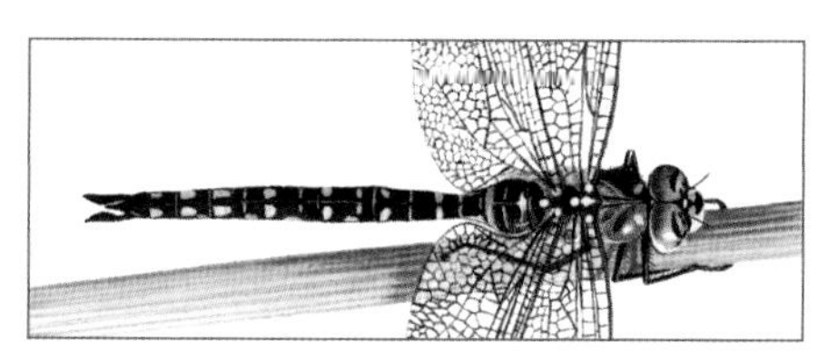

Migrant Hawker (63mm)
Key identification feature: T-shape on segment 2 of abdomen.
Flight period: August–October.

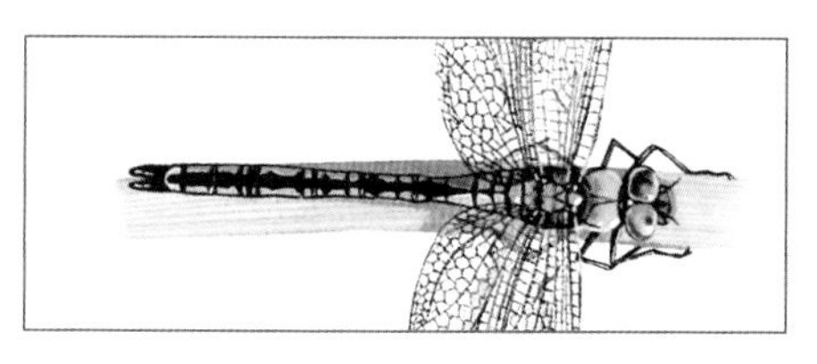

Emperor (78mm)
Key identification feature: green/blue-green eyes and apple-green thorax.
Flight period: June–August.

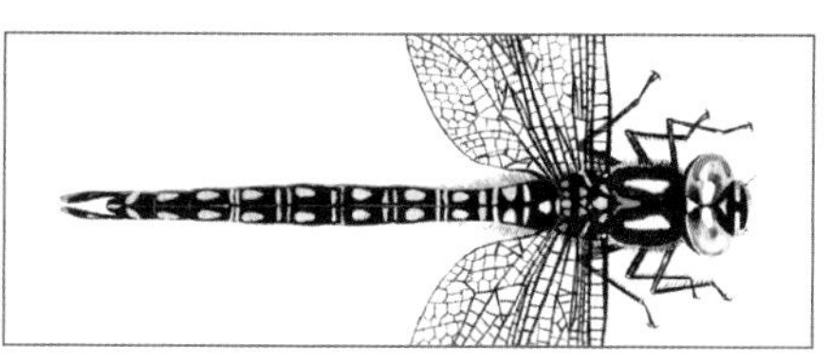

Hairy Dragonfly (55mm)
Key identification feature: hairy thorax.
Flight period: May–July.

Downy Emerald
Cordulia aenea Smaragaid Umha-dhaite

DESCRIPTION: *Male*: Length 48mm. Dark metallic green body with bright green eyes. Abdomen is club shaped, broadest near the tail and with short, slightly curved appendages. *Female*: Similar to male but stouter abdomen is less markedly widened toward the tail.
SEASON: Adults emerge during a few weeks in late May and the flight season extends to mid-July.
NATURE NOTES: Restricted to small, sheltered lakes in wooded or boggy areas in Kerry, west Cork and west Galway. When landing, it hangs from vegetation rather than perching. Adults hunt in woodland near breeding sites. Territorial males fly fast and low over water, challenging other dragonflies. They grasp females in their territory and mate nearby. Females then lay eggs alone, dipping the abdomen into water over submerged vegetation. Nymphs probably take two years to develop.
CONFUSION: The Northern Emerald is similar but the male abdomen is broadest near the middle and the appendages are long and calliper shaped.

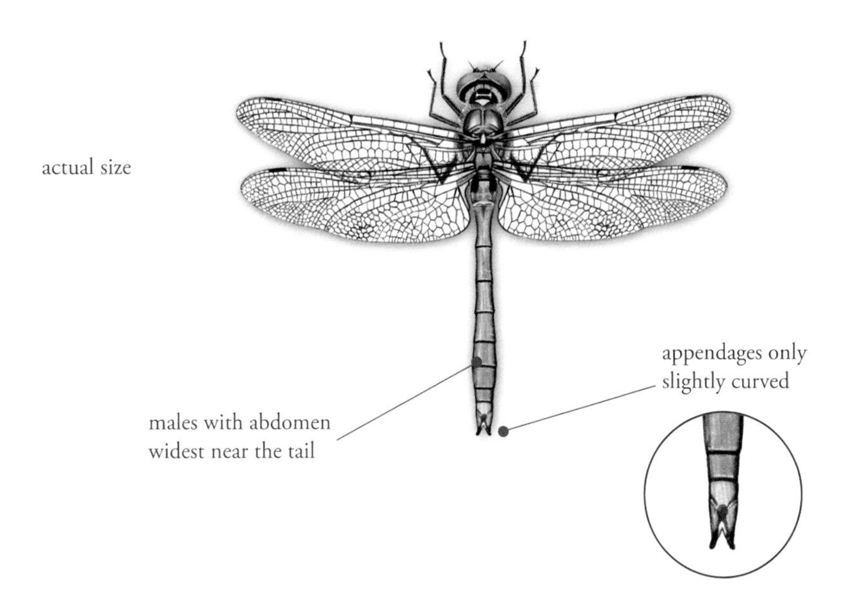

Northern Emerald
Somatochlora arctica
Smaragaid an Mhóintigh

DESCRIPTION: *Male*: Length 48mm. Dark metallic green body and bright green eyes. Abdomen is narrow at the base, broadest near the middle and with calliper-shaped appendages. *Female*: Similar colouration to male but with a stouter, unwaisted abdomen, two yellow spots on the third abdominal segment and smaller eyes.
SEASON: Emergence begins in late May. Flight season extends into August and possibly later but little is known of Irish populations.
NATURE NOTES: Very localised and known to breed at just a few sites in Killarney National Park with one sighting of an adult on Garnish Island in west Cork. Breeds in shallow, peaty bog pools and flushes. Adults feed near woodland margins. Territorial males patrol areas of potential breeding pools, flying low with periods of hovering. Once a female is encountered, the pair flies off to mate on a tree. Female lays eggs alone, dipping the upturned abdomen repeatedly in small pools with Sphagnum moss. Nymphs take at least two years to develop.
CONFUSION: Similar to the Downy Emerald but the abdomen is broadest near the middle and the anal appendages are different. Both species occur in Killarney National Park although they breed in different habitats.

actual size

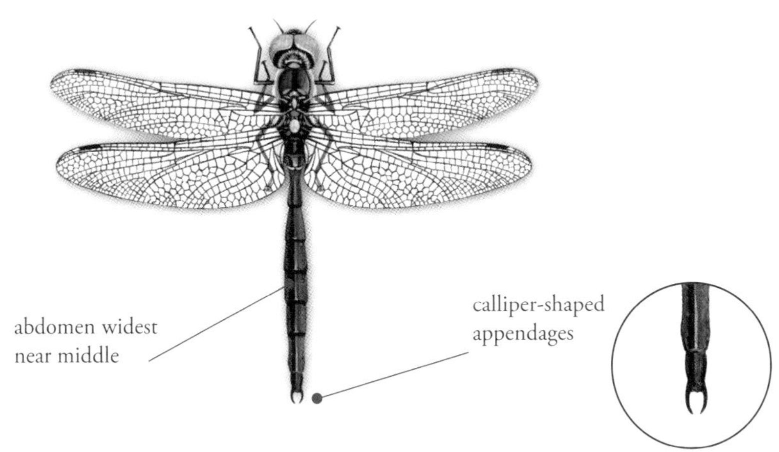

Four-spotted Chaser
Libellula quadrimaculata
Ruagaire Ceathairbhallach

Description: *Male*: Length 39–48mm. The thorax and abdomen are brown with the abdomen dark tipped and with yellow markings along each side. The wings are dark at the base and with dark spots on the leading edge. Occasionally specimens have dark patches near the wing tips. *Female*: Very similar to males.

Season: Emergence begins in early May and the flight season extends through to mid-August.

Nature notes: Common and widespread in most freshwater habitats but most abundant in bogs pools, fens and in sheltered margins of low-nutrient lakes. Males are territorial and aggressively defend territories, often from a prominent perch from where they can dash to investigate approaching dragonflies or prey. Mating is very brief and takes place on the wing. Males guard females as they lay eggs by dipping the abdomen into water. Nymphs take two years to develop.

Confusion: Readily recognisable by the shape, colouration and markings on the wings.

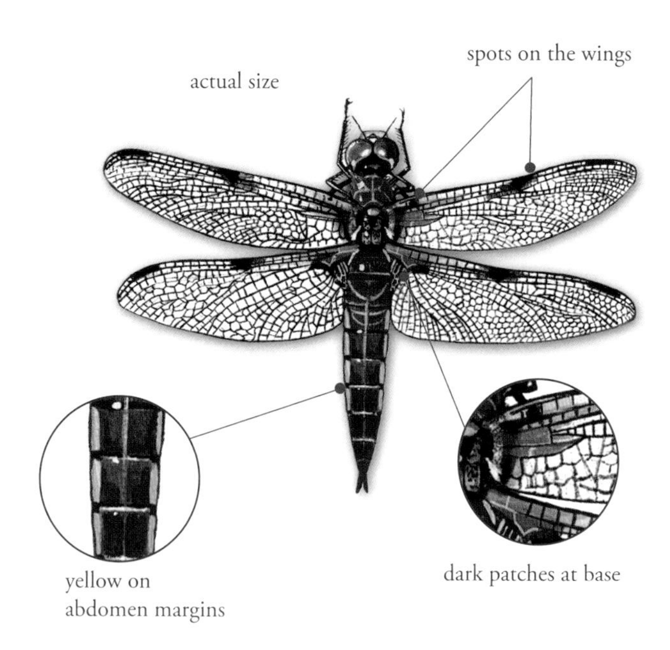

Black-tailed Skimmer
Orthetrum cancellatum
Scimire Earrdhubh

Description: *Male*: Length 44–49mm. Mature males have a powder-blue abdomen with a black tip and yellow oval spots on the sides. Wings are clear with a yellow costa. Immature males have a yellow abdomen with two longitudinal black bands on each segment that are gradually obscured as the blue colour develops. *Female*: Similar to immature males. Abdomen is yellow with two longitudinal black lines on each side. Older females become dark brown.
Season: Emergence begins in late May with peak numbers in June and July. Flight season ends in August.
Nature notes: Widespread but not common in central and western Ireland in base-rich limestone lakes. It also occurs in shallow ponds, including ponds on dunes and machair. Adults frequently bask on bare surfaces. Territorial males patrol an area of shoreline from a prominent perch. Females approach water to breed. Mating is very brief and takes place in flight. Females lay eggs in water by dipping the abdomen into it. Nymphs develop in shallow warm water over two years and live semi-buried in sediment.
Confusion: Male Keeled Skimmers are also blue on the abdomen but they are smaller and live in a different habitat. Females may be confused with Four-spotted Chaser and female darters but the black markings are distinctive.

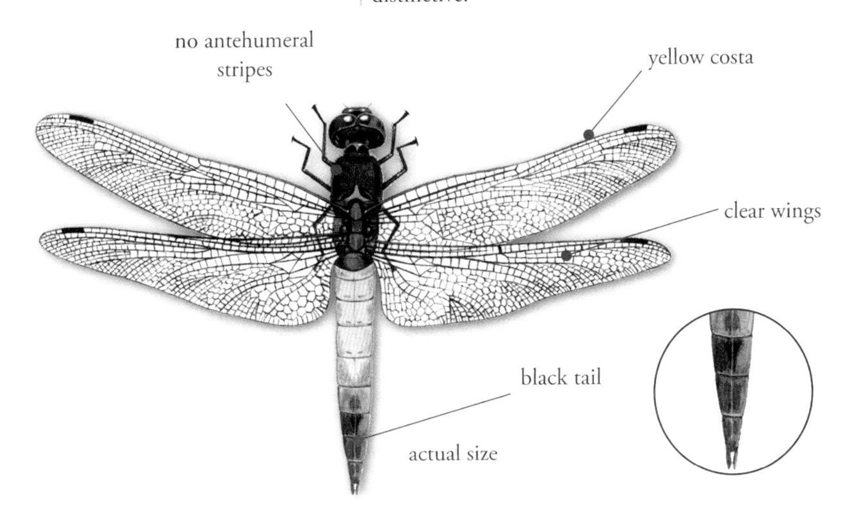

Keeled Skimmer
Orthetrum coerulescens Scimire na Sruthlán

DESCRIPTION: *Male*: Length 40–44mm. Mature males have a powder blue, narrow, tapered abdomen which is blue at the tip. Thorax has pale antehumeral stripes. Wings have a dusky tint near the tip. Immature males are golden yellow on the abdomen with brown markings but become blue as they mature. *Female*: Abdomen is yellow-brown with a thin, dark central stripe and darker markings near the sides. Wings may have small yellow patches near the base and some yellow veins.
SEASON: Emergence begins around the first week in June and the flight period extends to late August.
NATURE NOTES: Most common in southwestern and western Ireland but with scattered colonies across Ireland. Rare in the north. This species is restricted to flushes and shallow pools in heaths and bogs, mainly in upland areas. Males defend small territories around breeding sites and challenge passing dragonflies. Mating is brief and females lay eggs by flicking the tip of the abdomen in water, usually with the male in attendance. Nymphs take two years to develop.
CONFUSION: Male Black-tailed Skimmers are also blue on abdomen but have a black tail and live around limestone lakes and ponds. Female Black Darters have more black on the sides of the thorax.

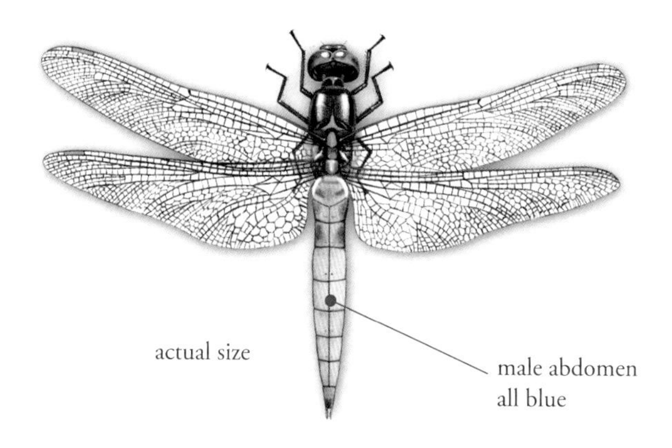

Common Darter

Sympetrum striolatum Sciobaire Coiteann

Description: *Male*: Length 40mm. Abdomen orange-red and unwaisted. Thorax brown above with yellow patches on the sides. Immature males are yellow-brown. Legs have a pale line on the tibiae. *Female*: Abdomen is yellow-brown, thorax light brown.

Season: Adults emerge from mid-June with peak abundance in mid-August. Flight season can extend into November in mild weather.

Nature notes: Common and widespread in most freshwater habitats but most common around ponds and lakes, especially where there is shallow water. Adults are very mobile and are often seen far from water, frequently perched on bare ground. Males are territorial and challenge passing darters. Females straying into their territory are seized and the pair flies off to mate. Eggs are laid in tandem with the eggs released into water. Larval development takes one year.

Confusion: Male Ruddy Darters have all-black legs and deeper red waisted abdomen. Females are most likely to cause confusion. Keeled Skimmer females are more strongly marked with black on abdomen and brown rather than yellow on the sides of the thorax. Two uncommon migrant species occasionally show up in Ireland: the Red-veined Darter which has red veins in the wings and the Yellow-winged Darter which has large yellow patches on the wings.

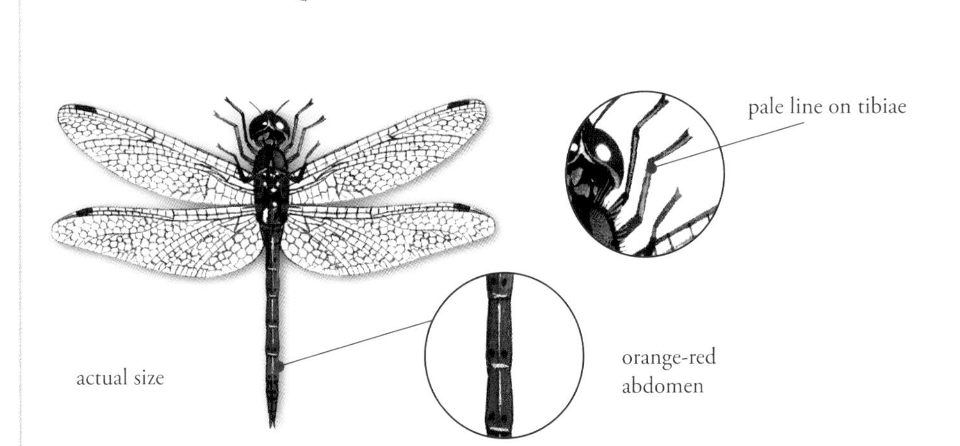

Ruddy Darter
Sympetrum sanguineum Sciobaire Cosdubh

DESCRIPTION: *Male*: Length 34mm. Mature males have a waisted and distinctly club-shaped abdomen. Immature males initially resemble females because of the yellow abdomen and brown thorax, but the red colouration develops over about 10 days as they mature. *Female*: Abdomen is unwaisted and golden-yellow with small black marks along each side.
SEASON: Emergence begins in mid-June with peak abundance in August. Flight season extends into mid-September.
NATURE NOTES: Common and widespread throughout the midlands and south in fens, cutover bogs, turloughs and small lakes although it avoids acidic water. Mating takes place perched near water. Eggs are laid while swooping over water in tandem, dipping tips of abdomen to shed eggs over submerged plants or mud around water margins. Eggs hatch shortly after, or, if late in the year, remain dormant until spring. Nymphs complete development by mid-summer.
CONFUSION: Mature males are distinctive although they may be overlooked amongst Common Darters. Immature males and females have all-black legs without any longitudinal stripe. Two uncommon migrant species occasionally occur in Ireland: the Red-veined Darter which has red veins in the wings and the Yellow-winged Darter which has large yellow patches on the wings.

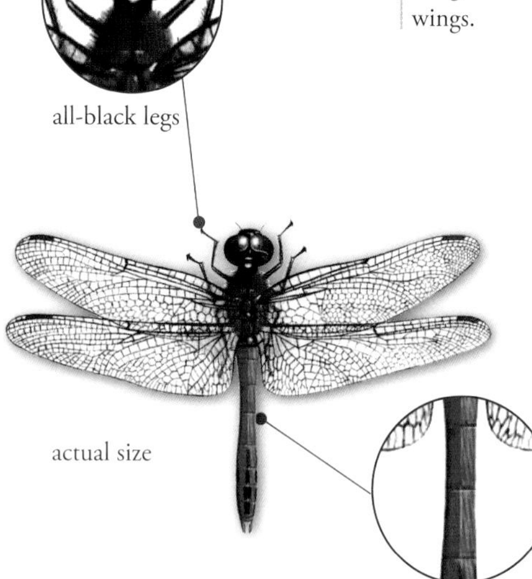

small patch of yellow on wings

waisted abdomen

Black Darter
Sympetrum danae Sciobaire Dubh

Description: *Male*: Length 32mm. Mature males are mostly black with two yellow stripes on thorax sides and yellow spots on the waisted abdomen. Eyes are pale underneath. Immature males are yellow on the abdomen with brown on the thorax. Legs are black. *Female*: Similar to immature male, mainly yellow on the upper surface of abdomen and black underneath. Legs are black. Dark triangle on front of thorax.

Season: Adults emerge around mid-July with peak numbers in August and September and are usually active until the end of October.

Nature notes: Widespread in the midlands, north and western seaboard. Absent from the southeast. Breeding habitat is acidic water in raised bogs, blanket bogs and nutrient-poor lakes. Males are not territorial and pairs arrive at breeding sites in tandem. Eggs are often laid in tandem, into water or Sphagnum moss. Eggs hatch the following spring and nymphs develop the following year.

Confusion: Mature males are distinctively black. Females could be confused with female Common Darters, Ruddy Darters and Keeled Skimmers. Female Black Darters have a more strongly marked thorax with a dark triangle on the upper surface and contrasting yellow and black sides.

actual size

Brown Hawker
Aeshna grandis Seabhcaí Omrach

DESCRIPTION: *Male*: Length 73mm. Thorax and abdomen mostly brown. The thorax has two yellow stripes on each side. The waisted abdomen has blue markings along the sides and the eyes are blue tinted. The wings are brown tinted with a brown costa. *Female*: Abdomen is unwaisted with yellow markings, eyes with a yellow tint.
SEASON: Adults begin to emerge about mid-June. Peak abundance is in mid-August and by mid-September numbers decline.
NATURE NOTES: Widespread and common in lowland ponds, small lakes and fens, usually where there is shelter from trees or shrubs, but absent from uplands and intact bogs. It is a tireless flier and has a distinctive flight of long glides interrupted with rapid, shallow wingbeats. Immature adults are common in woodlands. Males are less territorial than other hawkers. Females lay eggs alone, often into rotting wood. Nymphs take between two and four years to develop.
CONFUSION: Easy to recognise by the amber wings.

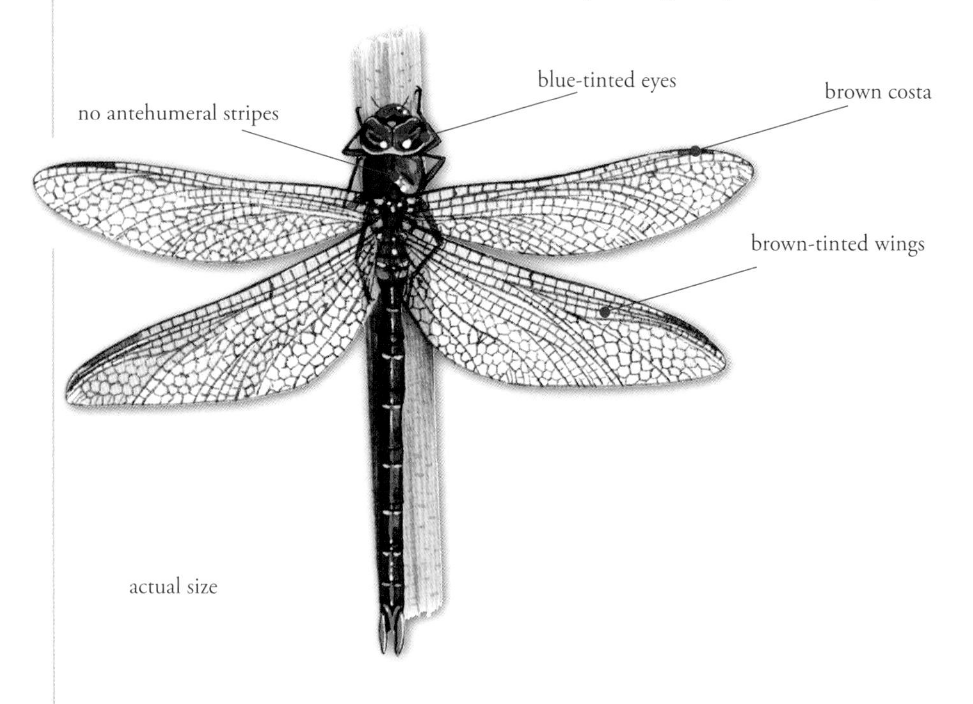

Common Hawker

Aeshna juncea Seabhcaí an Phortaigh

Description: *Male*: Length 74mm. Abdomen is dark with blue spots and yellow flashes. Antehumeral stripes are yellow and narrow. There are broad yellow stripes on the sides of the thorax. A diagnostic feature is the bright yellow costa. *Female*: Abdomen is brown with yellow (sometimes green) spots and flashes. Antehumeral stripes reduced to yellow spots.

Season: Adults begin to emerge as early as mid-June but peak abundance is in August. Flight season extends into October.

Nature notes: Common and widespread in Ireland. Breeds in a wide range of standing-water bodies from bog pools to large lakes and mountain loughs. It is most common in acidic water habitats but also lives in alkaline or neutral waters. Adults may occur far from water. After mating, egg laying takes place alone with eggs inserted into submerged plant stems. Larvae can take three to four years to develop, or even longer at altitude.

Confusion: The Brown Hawker has amber wings. The Hairy Hawker flies earlier in the year. The Migrant Hawker in the south and east is most likely to be confused but it has the yellow 'T' mark on the abdomen.

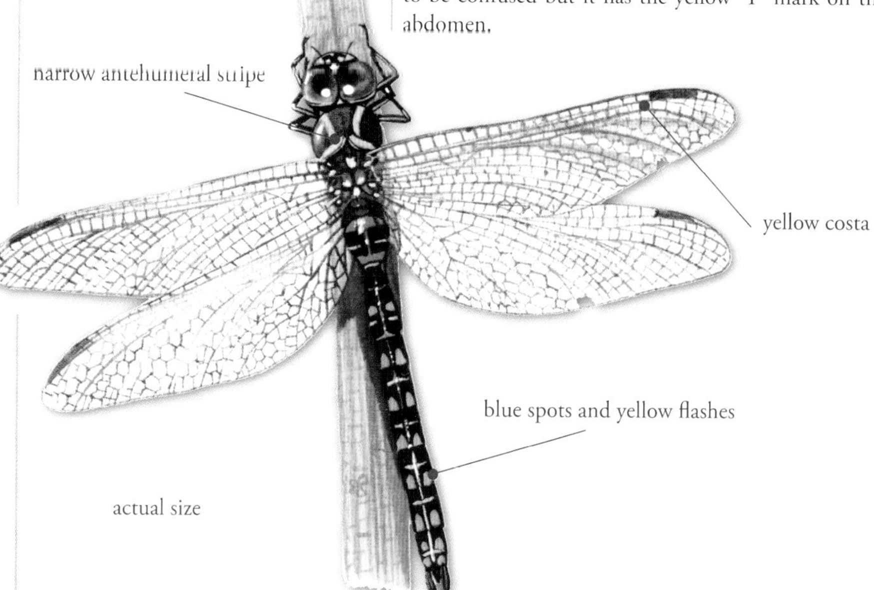

Migrant Hawker
Aeshna mixta Seabhcaí an Fhomhair

Description: *Male*: Length 63mm. Abdomen is dark with blue spots and a yellow T-shaped mark on the base. Antehumeral stripes are reduced to yellow spots. Immature adults have pale or lilac spots. *Female*: Similar to male but with yellow or yellow-green spots on the abdomen.
Season: Emergence begins in late July but flight season extends to late October.
Nature notes: Common in the south and east, having recently become established in Ireland as a breeding species. Breeds in lakes, ponds, large rivers and brackish water. Adults may be seen at woodland margins and hedgerows. Nymphs rapidly develop in warm, shallow water. Emergence occurs at night and immature adults stay away from water for about 10 days. Males patrol a shoreline territory and females are actively pursued. Mating takes place in flight. Egg laying is usually solitary, into stems of emergent plants or mud. Nymphs develop in one year.
Confusion: Most similar to the larger Common Hawker which has a darker thorax, yellow costa and lacks the T-shaped mark.

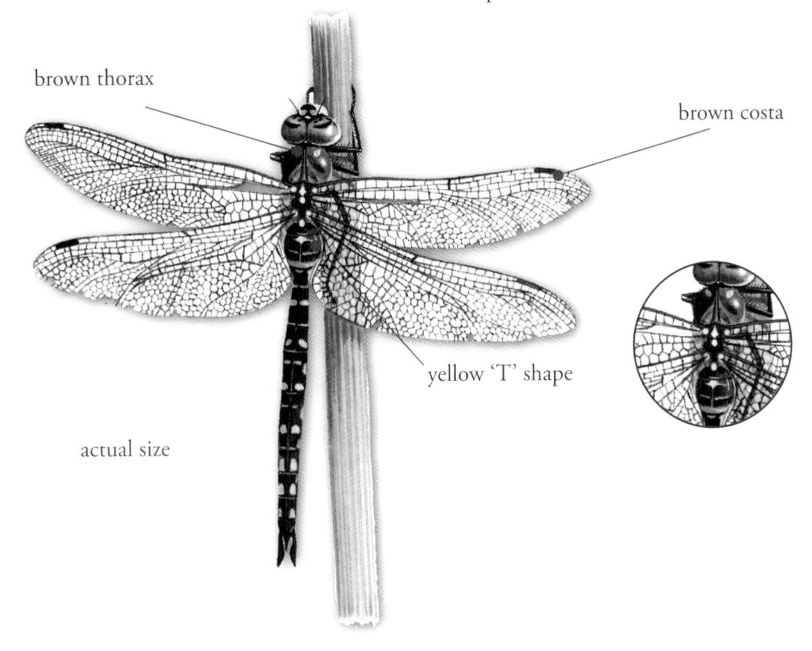

Emperor
Anax imperator Impire Gorm

Description: *Male*: Length 78mm. Thorax is bright green, eyes green and blue. Abdomen has striking blue markings and a black stripe on the midline. Abdomen usually held down-curved. Immature adults are greenish on abdomen. *Female*: Similar to males but with a green abdomen.
Season: Adults emerge from late May with peak abundance around mid-July. Adults are long-lived and may still be on the wing into late September.
Nature notes: Recently established as a breeding species in Ireland and now common in the south and east in well-vegetated natural and artificial ponds and lakes. Adults are active hunters over water, rarely settling and even consuming other dragonflies. Mating takes place while perched and the female lays eggs alone, inserting them into submerged plants. Nymphs live close to water surface and are initially striped, then turn greenish. Development takes two years or occasionally one year.
Confusion: The combination of bright green thorax and lack of spots on the abdomen make this species easily recognisable.

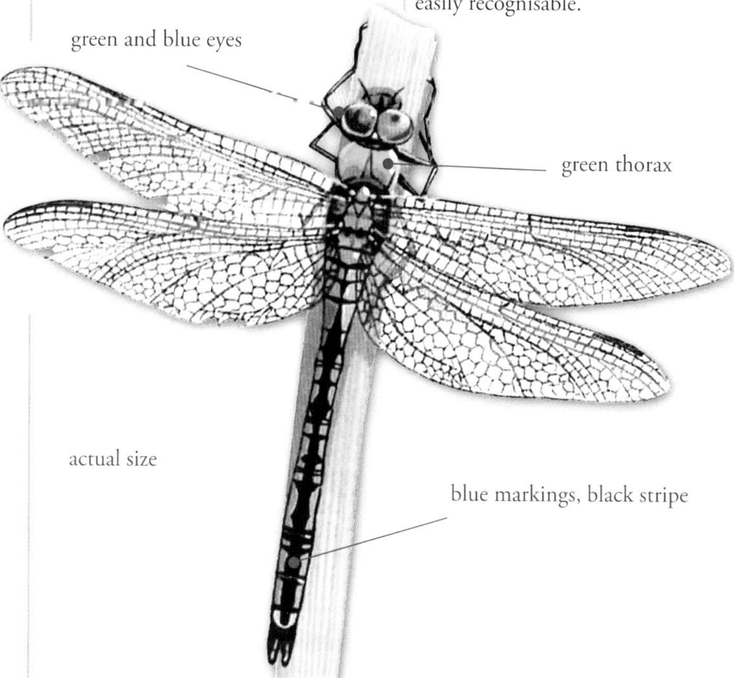

Hairy Hawker
Brachytron pratense Seabhcaí an Earraigh

Description: *Male*: Length 55mm. This is a small hawker with a dark brown abdomen and long anal appendages. The pear-shaped spots on dorsal surface are greenish-yellow and blue. The thorax is distinctly hairy. *Female*: Similar to males but with yellow spots and markings on the abdomen and antehumeral stripes reduced to yellow spots.
Season: Flight season is short, adults emerge from early May, peak abundance is around mid-June with only occasional sightings later than early July.
Nature notes: Widespread and locally common on sheltered lowland lakes, fens and cutover bogs, usually with fringing emergent vegetation. It is most frequent in the north midlands, Clare and around the south coast. Males are territorial. After mating, females lay eggs alone, inserting eggs into stems of pondweeds. Nymphs are thought to develop over two years.
Confusion: The early flight season makes this species difficult to confuse with other hawkers.

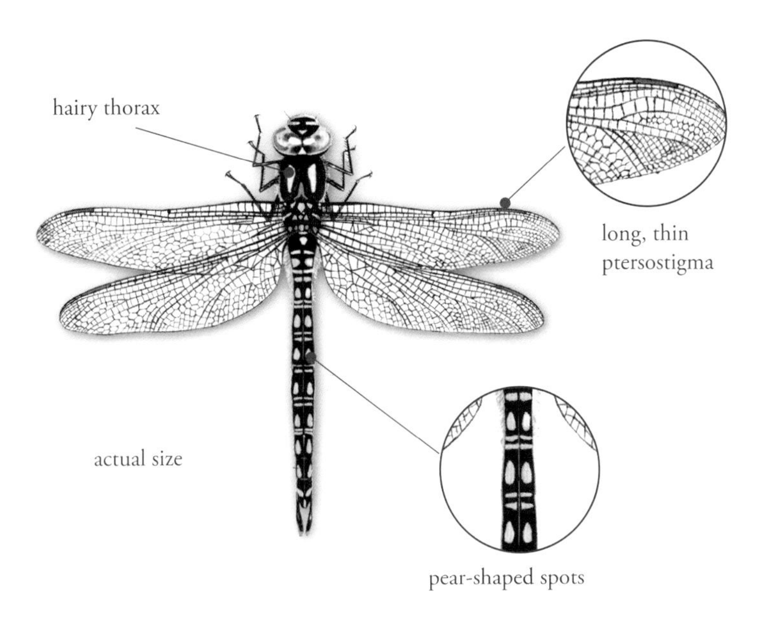

Bumblebees

Bees are closely related to wasps and ants and, together with sawflies and parasitic wasps, make up one of the largest orders of insects, the Hymenoptera. There are over 3,000 species of hymenopteran insect in Ireland: 100 are native bee species, of which 20 are bumblebees and the remaining 80 species are solitary bees. There is also the non-native domestic Honey Bee, which is familiar to everyone.

Similar to the Honey Bee, bumblebees are social insects but their colonies are much smaller, typically with 50 to 200 bees at peak population in mid- to late summer. A queen bumblebee establishes a colony in a nest on her own, unlike the Honey Bee which starts a colony via a swarm. Bumblebee nests are simple structures, in a cavity such as a mouse burrow, unlike the complex hive of the Honey Bee.

Bumblebees are important pollinators of a great variety of wild plants and commercial crops and are thus ecologically and economically an important group. However, populations of some species have declined significantly in Ireland in recent decades and six bumblebee species are threatened with extinction.

Life history

During the winter months queen bumblebees hibernate in burrows and other sheltered nooks. In spring, and sometimes very early in spring when there are still few flowers and temperatures are too low for other insects to be active, queens emerge from their winter shelter and prepare for the coming year. The previous summer, queens mated with one or several male bumblebees. Now they must feed to replenish energy stores and nourish developing eggs. They feed on pollen and nectar and seek suitable nesting places in burrows or small cavities, such as a mouse burrow or an abandoned bird's nest. Carder bees collect moss and comb it into well-insulated, ball-shaped nests, usually above ground. When a nest is ready, the queen moulds a miniature wax pot into which she regurgitates nectar to serve as a food supply while she incubates her young. She also collects pollen and gathers it into a large ball, about the size of her body. Onto this she lays her first batch of eggs which will develop in the nest. Queens incubate the nest and vibrate the powerful flight muscles within the thorax to generate heat to speed the development of the young larvae.

The first generation of workers emerges from the nest in late spring or summer and proceeds to collect pollen and nectar for subsequent generations of the colony. Workers tend the nest, make new cells for developing larvae and forage outside for pollen and nectar to nourish the developing young. Nests of some species have up to 200 workers but other species have fewer. From mid- to late summer some eggs develop into new queens. Workers may also lay unfertilised eggs and these develop into males. Once males and new queens have left the nest and mated, the function of the colony has been fulfilled and with the onset of cold weather in autumn and winter all except the hibernating queens die off.

Cuckoo Bumblebees

Cuckoo bumblebees employ an alternative strategy for survival and reproduction. They are close relatives of social bumblebees but, as the name suggests, they do not trouble themselves with nest building and developing a colony of their own; instead, females invade nests of social bumblebees and exploit the nest's resources. Cuckoo bumblebee females emerge from hibernation in late spring or early summer, later than social bumblebee queens. By this stage social bumblebee queens have established their nests. The cuckoo bumblebee female has similar colouration to queens of the host species that she seeks out. She invades a nest by force or by stealth, overpowering the queen and laying her own eggs. The existing workers continue to forage for food and tend to the young of the invader. Cuckoo bumblebees do not produce workers, only females and males which leave the nest once they are adults.

Heath Bumblebee female

Heath Bumblebee queen making a wax pot to store food while she incubates the first generation in a new nest.

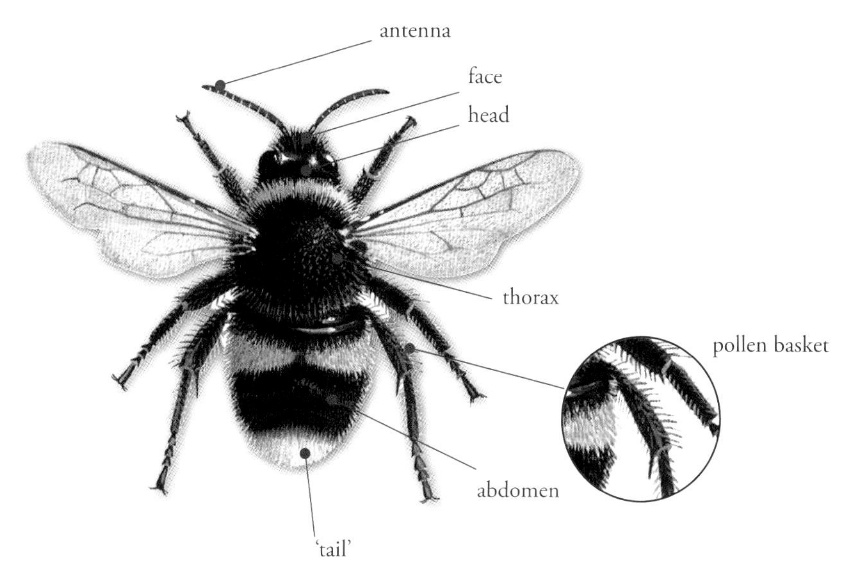

Bumblebee body parts useful for identification.

Anatomy and identification

Most of the 20 species of bumblebee in Ireland are easily identified with a little effort. The first feature to note is the tail colour. Is it white, red, ginger or yellow? Then inspect the colours on the rest of the body. Is it uniform or are there bands of yellow, orange or white? This will help identify 17 of the 20 Irish species. Three species of 'White tailed Bumblebee' are so closely similar that they are not readily distinguishable without genetic analysis.

Queens can usually be identified by their very large size. Workers are smaller and early in summer some workers are much smaller than queens. Males of some species can be difficult to distinguish from females but some species have distinct differences in colouration between the sexes. Also, males do not have pollen baskets on their hind legs and their behaviour is somewhat lethargic in comparison to the usually busy queens and workers. Pollen baskets are visible as the shiny outer face on the hind leg, fringed with long hairs where pollen is carried. Cuckoo bumblebees can usually be identified by the sparse hairs on the body, darkened wings and lack of pollen baskets.

Note on maps of bumblebee distribution in Ireland

Maps for each species highlight vice-counties with records since 1990. Several species of bumblebee have undergone sharp declines in abundance and range in Ireland over recent decades, in particular the Shrill Carder Bee and the Great Yellow Bumblebee. Declines in populations of these bumblebee species are likely to be due to changes in agricultural practices leading to decreased abundance of flowers – bumblebees' primary food source in the countryside.

White-tailed Bumblebees

One yellow thorax band

Buff-tailed Bumblebee

White-tailed Bumblebee

Forest Cuckoo Bumblebee

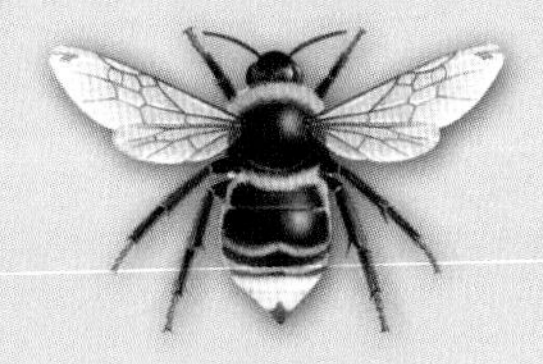

Gypsy Cuckoo Bumblebee

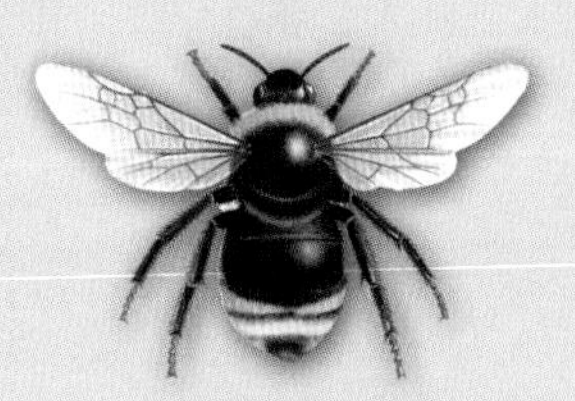

Southern Cuckoo Bumblebee

Two yellow thorax bands

Garden Bumblebee

Heath Bumblebee

Barbut's Cuckoo Bumblebee

Red-tailed Bumblebees

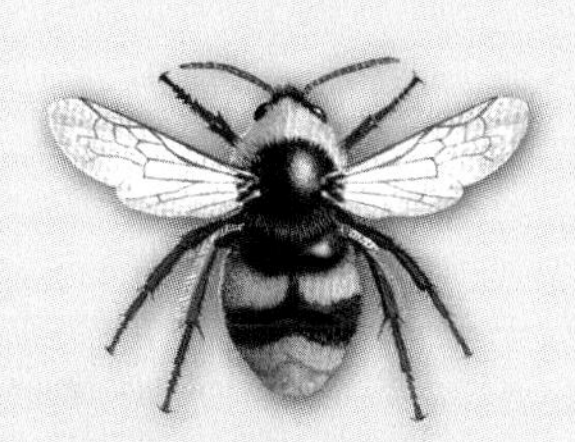

Early Bumblebee

Red-tailed Bumblebee

Red-shanked Carder Bee

Red-tailed Cuckoo Bumblebee

Mountain Bumblebee

Shrill Carder Bee

Ginger or Yellow-tailed Bumblebees

Large Carder Bee

Common Carder Bee

Great Yellow Bumblebee

Field Cuckoo Bumblebee

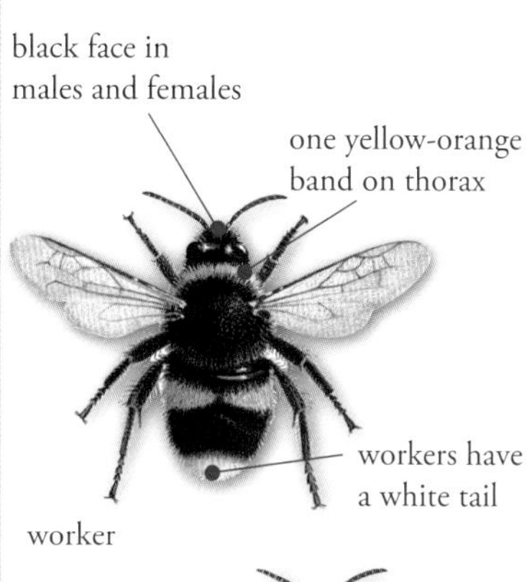

worker

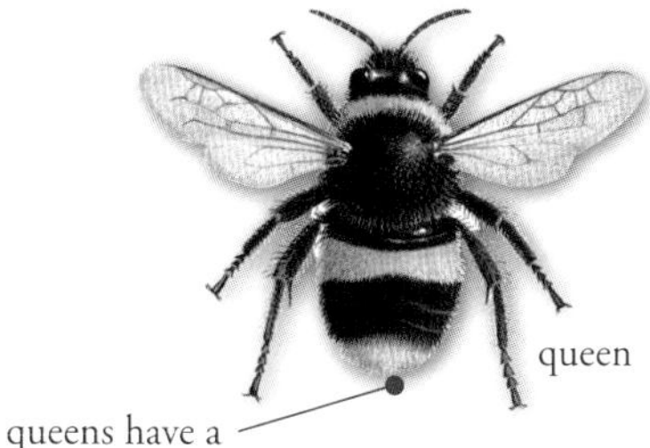

queen

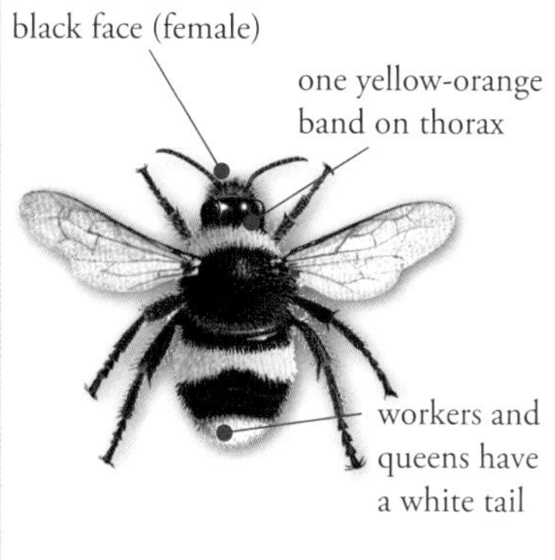

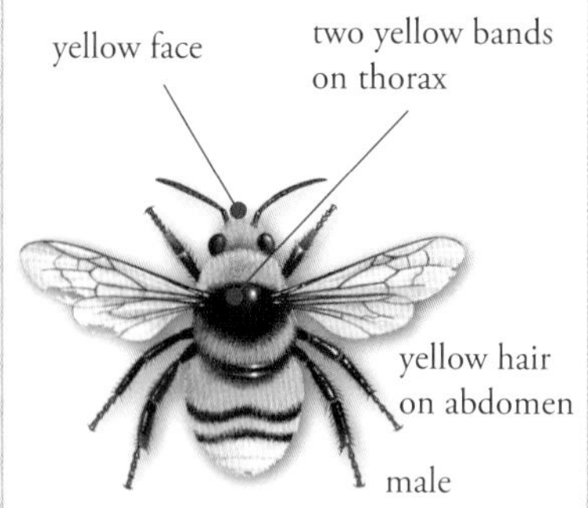

male

Buff-tailed Bumblebee *Bombus terrestris*

Description: *Queens and workers*: Yellow-orange bands on thorax and abdomen. Queens with tail grey/buff or pale orange. Workers have a white tail. Queens are large, workers may be much smaller. Face and the tongue are short. *Male*: Similar to workers.
Season: Queens emerge from February on, workers appear a few weeks later. Males appear from late July.
Nature notes: Widespread and common everywhere. Nests are made underground and can have up to 200 individuals. Visits a wide variety of wildflowers. Also an important pollinator of commercial crops and non-native populations are imported for this purpose.

Confusion: Workers are indistinguishable from White-tailed Bumblebee workers. Queens can be identified by the grey to buff tail and the darker yellow-orange bands. Males can be distinguished from White-tailed Bumblebee males by the black face.

White-tailed Bumblebee *Bombus lucorum* agg.

Description: *Queens and workers*: Bright yellow bands on the thorax and abdomen. Tail white. Face and tongue are short. *Male*: Face and front of thorax with fluffy yellow hairs.
Season: Queens emerge from February, workers appear in early summer. Males appear around July.
Nature notes: Common and widespread in gardens, bogs and coastal habitats. These species visit a wide range of flowers and may rob nectar from tube-shaped flowers. Three very closely similar species occur here: the White-tailed Bumblebee (*Bombus lucorum* (*sensu stricto*)), the Northern White-tailed Bumblebee (*Bombus magnus*) and the Cryptic Bumblebee (*Bombus cryptarum*).
Confusion: Queens can usually be easily distinguished from Buff-tailed Bumblebee queens by the white tail. Workers cannot be reliably distinguished in the field. Males have yellow hair on the face and thorax, unlike male Buff-tailed Bumblebees.

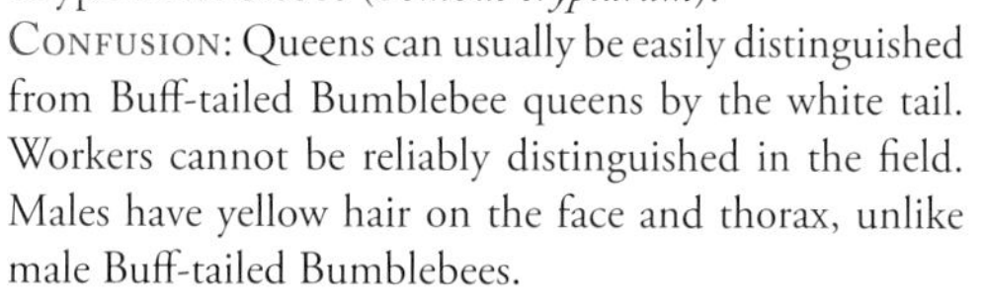

Forest Cuckoo Bumblebee *Bombus sylvestris*

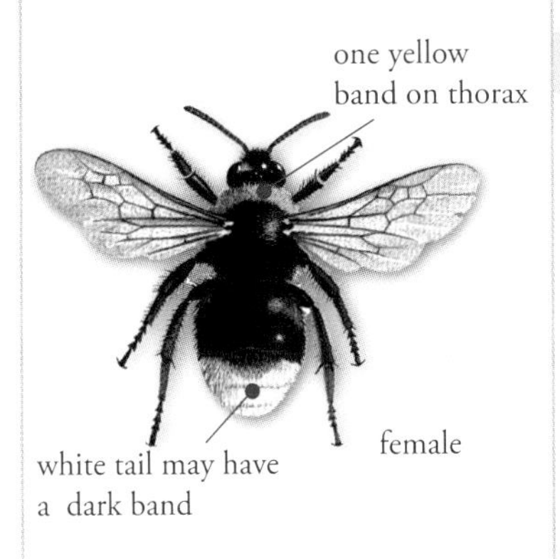

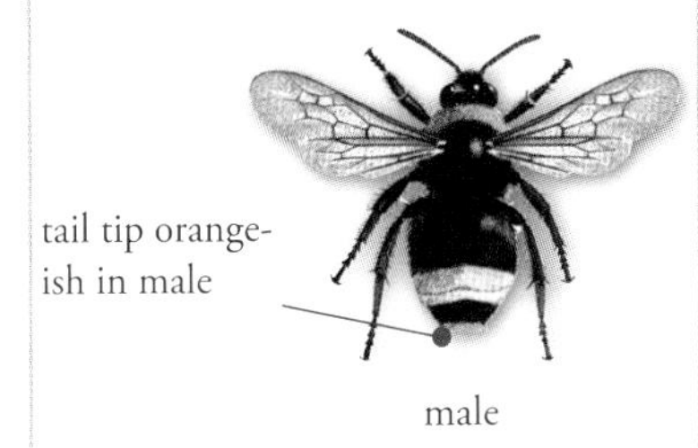

DESCRIPTION: *Female*: Up to 16mm in length. Thorax with one yellow band. White tail curled underneath. Wings darkened. Face and tongue short. Sometimes a second, faint yellow band on the thorax. Tail white or yellowish. *Male*: Black band near the tip of the white tail and orange hairs on the very tip. Face hair black.
SEASON: Females emerge in April, males appear from May onwards and both fly until October.
NATURE NOTES: The most widespread and frequently recorded cuckoo bumblebee in Ireland. It invades nests of the Early Bumblebee and Heath Bumblebee. It visits a range of flowers including dandelions, scabiouses, knapweeds and blackberry.
CONFUSION: Specimens with yellowish tails resemble the Field Cuckoo Bumblebee. That species has black hairs at the tip of the tail whereas those of the Forest Cuckoo Bumblebee are orange.

Gypsy Cuckoo Bumblebee *Bombus bohemicus*

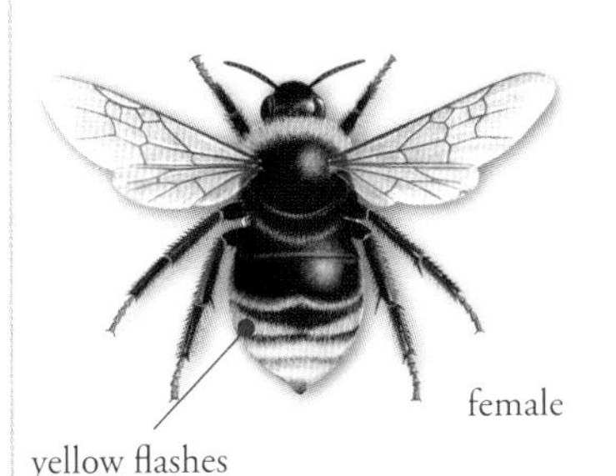

DESCRIPTION: *Female*: The thorax has one yellow band and the tail is white with narrow yellow flashes at each side. Face and tongue short. Wings are darkened. *Male*: Sometimes an additional pale yellow band on rear of thorax and/or at front of abdomen. There are black hairs on the face.
SEASON: Females emerge in April, males appear in May and both sexes may be on the wing until September.
NATURE NOTES: Widespread but uncommon in the habitat of the host, the White-tailed Bumblebee. It visits a range of flowers, including dandelions, scabiouses, vetches, knapweeds and thistles.
CONFUSION: This species is very similar to the extremely rare and slightly larger Southern Cuckoo Bumblebee, which has larger yellow flashes in front of the white tail. Microscopic examination is necessary to accurately distinguish these species.

female

Southern Cuckoo Bumblebee *Bombus vestalis*

DESCRIPTION: *Female*: One dark yellow band on thorax and yellow flashes at the front of the white tail. The face is as long as it is wide and the tongue is short. Large, up to 21mm. Wings may be quite dark. *Male*: Similar to female. May have a faint yellow band on front of abdomen. Face hair is black.

SEASON: Females emerge as early as April while males appear around June. This species is active until the end of September.

NATURE NOTES: A very rare species in Ireland with only a few old records from Wexford and Kilkenny from the 1920s. Refound in Ireland in 2014 in Co. Dublin. Invades nests of the Buff-tailed Bumblebee.

CONFUSION: Very similar to and hard to separate from the Gypsy Cuckoo Bumblebee.

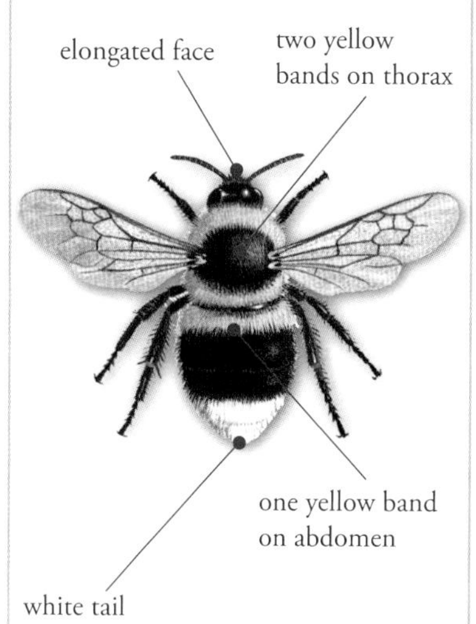

Garden Bumblebee *Bombus hortorum*

DESCRIPTION: *Queens and workers*: Two yellow bands on thorax and a yellow band on front of abdomen. The tail is white. The face is much longer than it is wide and the tongue is very long. Queens are large but workers vary in size. *Male*: Similar in colouration to females and with black hairs on the face.

SEASON: Queens emerge in March or April, workers appear around the end of May. Males appear from June onwards. They may still be on the wing in October.

NATURE NOTES: Very common and widespread in Ireland. Frequent in gardens but also in most other habitats. It visits a wide range of often tube-shaped flowers.

CONFUSION: Female Heath Bumblebees have similar colouration but the face is only about as long as it is wide in contrast to long-faced Garden Bumblebees. Males differ in the face colour and shape. Barbut's Cuckoo Bumblebee can look similar but it is less brightly coloured and has darkened wings, sparser hairs, short face, and females lack pollen baskets.

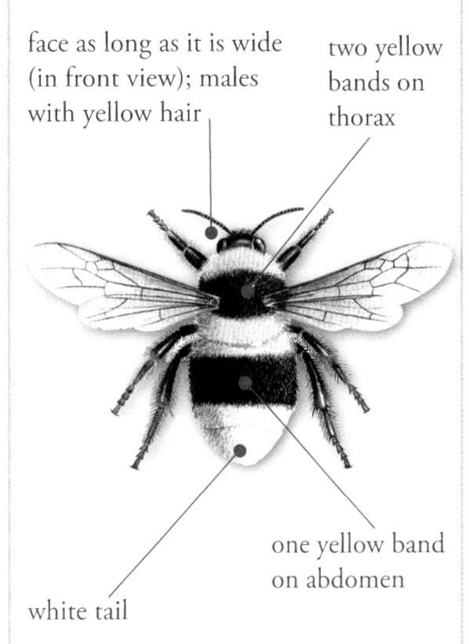

Heath Bumblebee *Bombus jonellus*

Description: *Queens and workers*: The thorax has two yellow bands and there is a yellow band on abdomen and the tail is white. The face and tongue are short. Queens and workers are smallish. *Male*: Similar to females but with yellow hairs on the face.
Season: Queens emerge in March, workers appear around June, males emerge some weeks later and fly until October.
Nature notes: Widespread and common particularly in bogs or heaths but also in gardens. Nests can be above or below ground and are small. Fond of heathers and other flowers and willows in spring.
Confusion: The Garden Bumblebee is generally larger and males do not have yellow hairs on the face which is elongated. Barbut's Cuckoo Bumblebee is usually less brightly coloured and has darkened wings, sparser hairs, and females lack pollen baskets.

Barbut's Cuckoo Bumblebee *Bombus barbutellus*

Description: *Female*: Two yellow bands on the thorax and sometimes a yellow band on the abdomen. The tail is white. The face as long as it is wide and the tongue is short. It is a large species. *Male*: Similar to females. Face hair colour is black.
Season: Females emerge around April. Males appear in June.
Nature notes: This is an uncommon species in Ireland with records mainly from the west and north. It feeds on a wide range of flowers in a variety of habitats. Invades nests of Garden Bumblebees but it is much less common than its host.
Confusion: Colouration is similar to both the Garden Bumblebee and Heath Bumblebee but Barbut's Cuckoo Bumblebee is less brightly coloured, less densely hairy, with tinted wings and lacks pollen baskets.

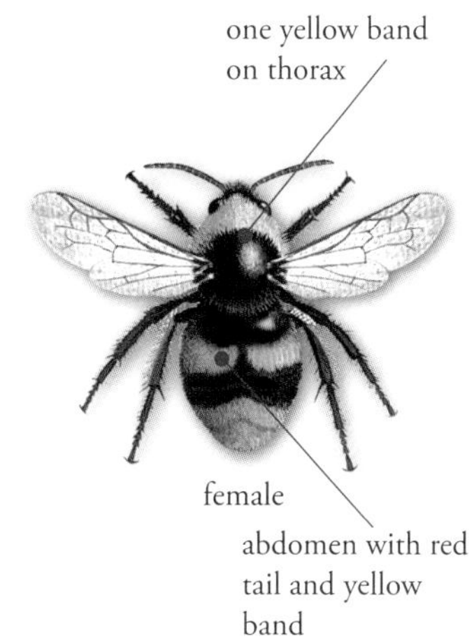

Early Bumblebee *Bombus pratorum*

Description: *Queens and workers*: Easily recognised by the orange tail and yellow bands on the abdomen and thorax. The face and tongue are short. Some workers are very small and dark and may be hard to recognise. *Male*: More yellow than female, especially the thorax and face. There may be a yellow band on rear of thorax.

Season: Queens emerge early, in February or March. Workers appear by May. Males appear in June. This species is active until October.

Nature notes: Common and widespread in many habitats, including gardens. Queens make nests underground or in holes in trees. Colonies are small, with fewer than 100 workers. Visits a wide range of both shallow and deep flowers.

Confusion: Male Red-tailed Bumblebees lack yellow on the abdomen. Male Mountain Bumblebees have more red hair on the abdomen.

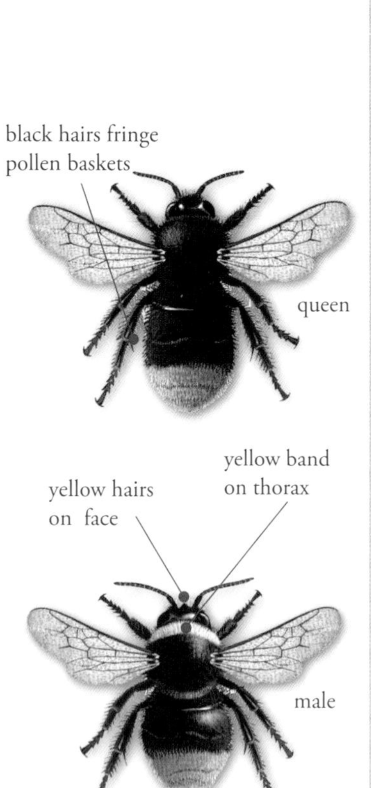

Red-tailed Bumblebee *Bombus lapidarius*

Description: *Queens and workers*: Black hairs on head, thorax and abdomen with a red-orange tail. Pollen baskets fringed with black hairs. The face is about as long as it is wide and the tongue is medium length. Queens are large, up to 20mm, but workers vary in size and may be quite small. *Male*: Bright yellow hairs on the face and usually a strong yellow band at front of thorax. There may also be a faint yellow band at rear of thorax. There are red hairs on the hind legs.

Season: Queens emerge from March onwards. Workers appear from May and males appear around June. Queens hibernate from October onwards.

Nature notes: Very widespread and common in Ireland in a variety of habitats including gardens and parks.

Confusion: Females are very similar to the Red-shanked Carder Bee but do not have the fringe of red hairs on the hind legs. Male Red-shanked Carder Bees have much less yellow on the thorax and face. Female Red-tailed Cuckoo Bumblebees have sparse hair on the body and darkened wings.

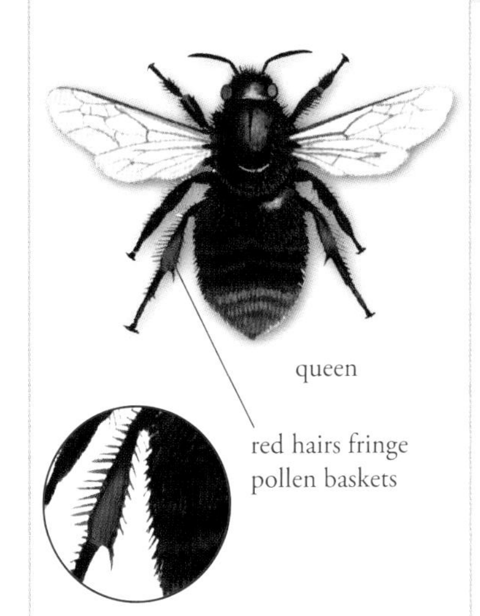

Red-shanked Carder Bee *Bombus ruderarius*

DESCRIPTION: *Queens and workers*: Head, thorax and abdomen black. Tail red-orange. Pollen baskets fringed with reddish hairs. Face elongated and tongue long. *Male*: Mainly black with a red-orange tail and with faint bands on the thorax. Some yellow hairs on the face.

SEASON: Queens emerge in April or May, workers a few weeks later and males emerge in July.

NATURE NOTES: Rare but locally common in flower-rich meadows and coastal habitats. Queens use moss to make a nest in a burrow or tussock. Visits a variety of flowers but favours clovers and knapweeds.

CONFUSION: Female Red-tailed Bumblebees have black hairs fringing the pollen baskets. Red-tailed Cuckoo Bumblebees are larger with darkened wings and are less densely hairy. Males have much less yellow hair on the face and thorax than the Red-tailed Bumblebee

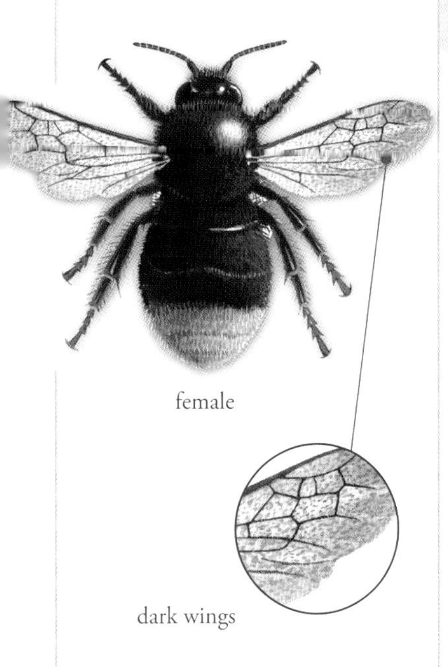

Red-tailed Cuckoo Bumblebee *Bombus rupestris*

DESCRIPTION: *Female*: Very large, up to 22mm long. Black all over except for the red tail. Wings are very dark. The face is as long as it is wide and the tongue is short. *Male*: There may be faint traces of pale bands on the thorax; face hair is black.

SEASON: Females emerge from May onwards. Males appear from June. This species can be on the wing into September.

NATURE NOTES: Uncommon in Ireland with most records coming from near the coast. Invades nests of the Red-tailed Bumblebee and Red-shanked Carder Bee. Visits a range of flowers, including dandelions, scabiouses, knapweeds and clovers.

CONFUSION: Queens of Red-tailed Bumblebees and Red-shanked Carder Bees are similar in colouration but they are smaller, more densely hairy, lack darkened wings and have pollen baskets. Males could be confused with male Red-shanked Carder Bees but they too lack the darkened wings and are more densely hairy.

Mountain Bumblebee *Bombus monticola*

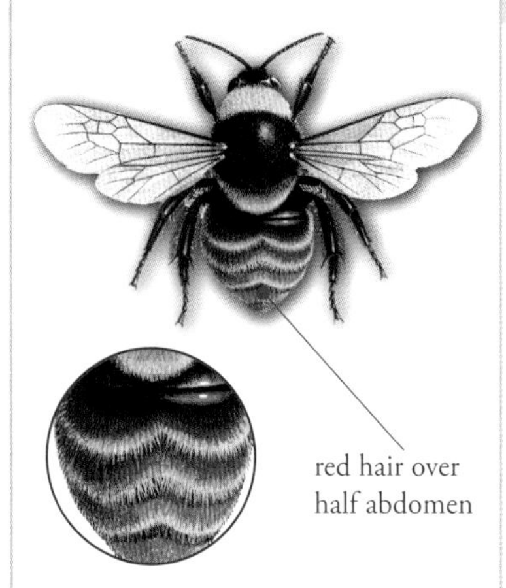

Description: *Queens and workers*: Strikingly red-orange tail extends over half the abdomen. Two lemon-yellow bands on thorax; the rear one may be quite narrow. The face is as long as it is wide and the tongue is short. It is a relatively small species. *Male*: Obvious yellow hair on the face but otherwise similar to female.
Season: Queens emerge in April. Males emerge from July onwards. This species can be on the wing into October.
Nature notes: First recorded in Ireland in 1974 in County Wicklow. It has since been recorded widely in uplands and coastal habitats in southeastern and northeastern Ireland. Nests above ground under heather. Visits mainly flowers of bilberry (Fraochan), heathers, bird's foot trefoil and kidney vetch.
Confusion: Other red-tailed species can be confused initially but the extensive area of red-orange hair on the abdomen should make this species quite obvious.

Shrill Carder Bee *Bombus sylvarum*

Description: *Queens and workers*: This distinctive bumblebee has two bands of grey-green hairs on the thorax with a dark band across the middle. The abdomen has grey-green hairs and an orange-red tail. The face is longer than it is wide and the tongue is long. It is medium sized, queens are up to 17mm in length. *Male*: Similar to females.
Season: Queens emerge in May. Males appear from July onwards and it overwinters from October.

Nature notes: This is a very localised species in Ireland with the main populations found in Clare and smaller populations in good quality, flower-rich habitats. Nests above ground amongst vegetation. Visits a wide range of flowers with a preference for tube-shaped flowers.
Confusion: It is the only grey-green bumblebee but worn individuals of the Common Carder Bee could look similar if they are faded or covered in pollen.

females have black hair
on the face

ginger orange
thorax

variable amounts of black
and grey hairs here

Common Carder Bee *Bombus pascuorum*

DESCRIPTION: *Queens and workers*: Thorax covered in orange hair, abdomen with a variable mix of grey, blonde, orange and black hair. Face elongated and the tongue is long. Medium sized, up to 17mm. Workers may be much smaller. *Male*: Similar to female but with orange hairs on the face.
SEASON: Queens emerge from overwintering as early as March, workers appear a few weeks later. Males appear around July.
NATURE NOTES: Very widespread and common everywhere. Nests are made from moss, usually at ground level. Visits a wide range of flowers with some preference for the pea family.
CONFUSION: The Large Carder Bee is more densely hairy and velvety and rather more orange. Old or worn Large Carder Bees may look like unusual versions of this species.

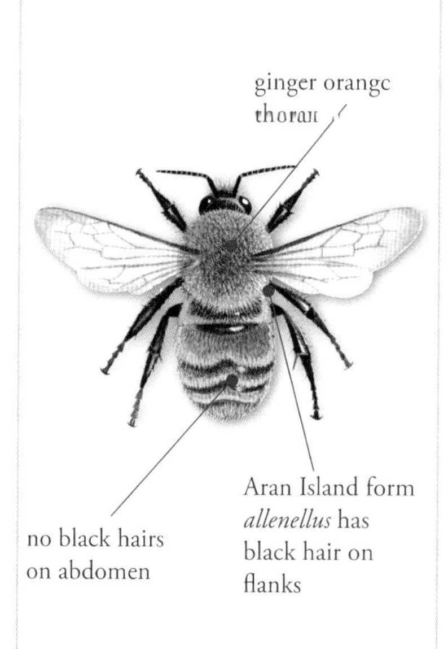

Large Carder Bee *Bombus muscorum*

DESCRIPTION: *Queens and workers*: The body is covered in dense orange hair. Thorax is a deeper orange than the rest of the body. The face is longer than it is wide and the tongue is long. *Male*: Similar to females.
SEASON: Queens emerge as early as March. Males appear around June. Recorded into September.
NATURE NOTES: Widespread across Ireland but becoming less common. Many records are from coastal areas with abundant flowers. Builds a nest of moss above ground in vegetation. Visits a variety of flowers with a preference for flowers of the pea family. There is a colour form endemic to the Aran Islands (var. *allenellus*) that has black hairs on the sides of the thorax.
CONFUSION: Pale specimens of the Common Carder Bee are similar but that species always has some black hair on the abdomen. Very similar to the Brown-banded Carder Bee (*Bombus humilis*), which has yet to be seen in Ireland.

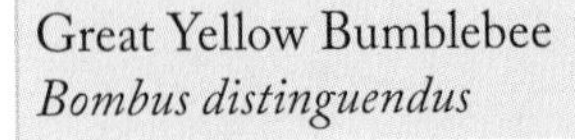

Great Yellow Bumblebee
Bombus distinguendus

mainly yellow with distinct black band on thorax

Description: *Queens and workers*: Mostly pale yellow on the thorax and abdomen but with a distinct black band across the thorax. The face is longer than it is wide and the tongue is long. Queens are large. *Male*: Similar to females.
Season: Queens emerge around May and workers a few weeks later. Males appear in July and August.
Nature notes: Formerly much more widespread, it is now very localised in Ireland. It has recently been found in only a few flower-rich habitats near the west coast and on esker ridges in central Ireland. It nests underground or under cover and it visits primarily clovers, knapweeds and thistles.
Confusion: Not easily confused with other Irish species but it could look similar to worn or faded specimens of the Large Carder Bee.

Field Cuckoo Bumblebee *Bombus campestris*

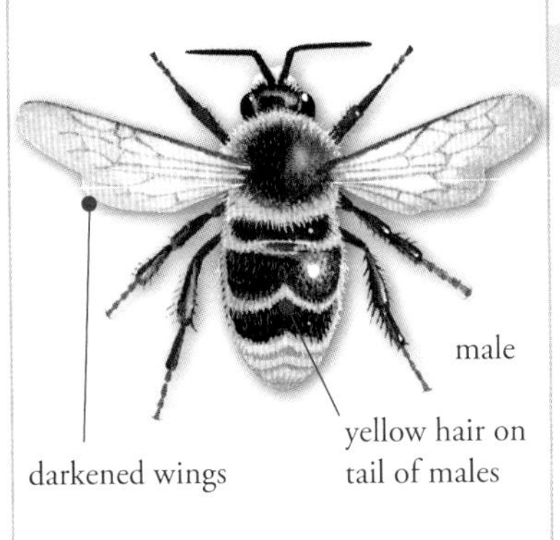

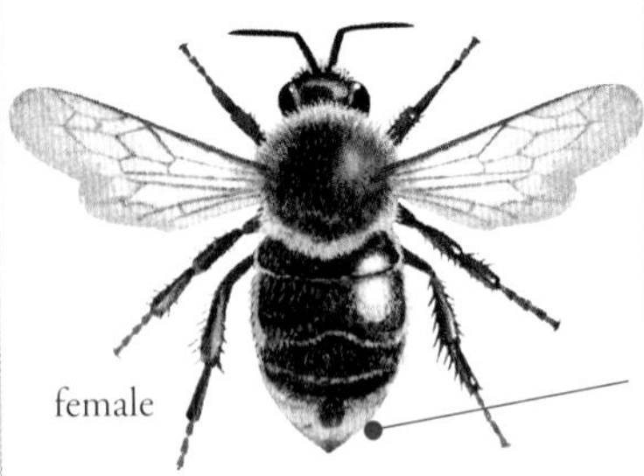

Description: *Female*: Two yellow stripes on thorax. Patches of yellow hairs on each side of abdomen. The amount of hair on the abdomen varies and specimens can be quite dark with little yellow hair. The face is as long as it is wide and the tongue is short. Females are up to 18mm in length. *Male*: Variable; the thorax may be all black or with one or two bands. Face hair is black.
Season: Females emerge from April onwards. Males emerge around June or later.
Nature notes: Widespread but uncommon in Ireland. Invades nests of the Common Carder Bee and Large Carder Bee. Feeds on a range of flowers, including dandelions, scabiouses, vetches and knapweeds.

Confusion: Similar to specimens of the Forest Cuckoo Bumblebee that have yellowish tails.

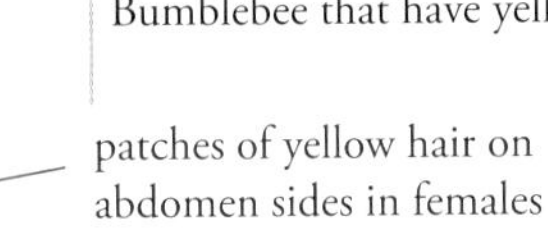

patches of yellow hair on abdomen sides in females

Grasshoppers and allies

This group of large or medium-sized insects is easily recognised by the large hind legs. The name Orthoptera, however, is derived from the Greek '*ortho*' meaning straight and '*ptera*' meaning wing, and refers to the parallel-sided structure of the front wings. There are over 26,000 species worldwide but only 13 species are resident in Ireland. The cool, damp Irish climate does not suit many of these warmth-loving species, most of which are restricted to warm, sheltered habitats. Some of Irish species are inhabitants of bogs and wetlands, particularly the Large Marsh Grasshopper, which is very rare in Britain but thrives in the wettest parts of raised and blanket bogs.

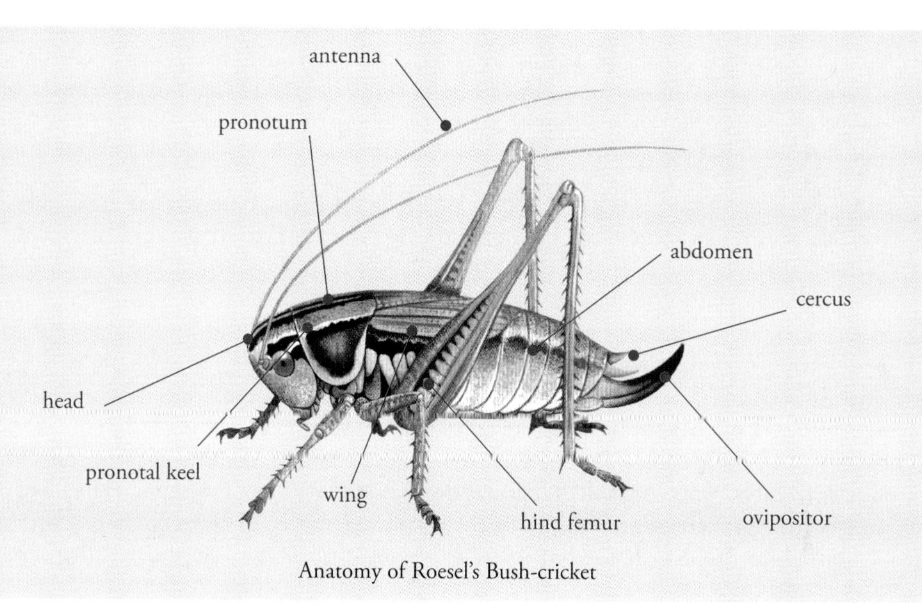

Anatomy of Roesel's Bush-cricket

Anatomy

The head has compound eyes with good vision for avoiding predators and judging distances. Strong mandibles on the underside of the head are capable of chewing tough plant material. Antennae can be long and filamentous or short and segmented. The pronotum is saddle-shaped and may have keels or ridges on the sides or midline. The forewings are tough and leathery and protect the more delicate membranous hind wings. In many species the wings are small or reduced to flaps. The hind femur is enlarged to accommodate powerful jumping muscles. Appendages on the hind abdominal segments are called cerci; in some species these are modified for grasping. Females have an egg-laying appendage called an ovipositor that is used to insert eggs into plants or soil. In many grasshoppers it is small and hidden within the abdomen but in bush-crickets it is very large and obvious.

Life cycle – how grasshoppers and allied insects develop

The egg is the overwintering stage for most species of grasshoppers and bush-crickets. The eggs of bush-crickets are always laid singly and inserted into stems or bark of living plants with the long ovipositor. Eggs of grasshoppers are laid in groups into soil and may be protected by a tough case or pod secreted by the mother. Eggs remain dormant during winter and nymphs hatch in the spring when temperatures rise and food becomes available. Most grasshoppers have four nymph stages and bush-crickets have five or six. Development takes about two months and the insects usually mature during summer. Metamorphosis is incomplete and there is no pupal stage.

Grasshoppers produce different types of song during courtship and mating. The songs can vary in intensity of sound, form of pulse and frequency of repetition. Single males or groups of males produce 'normal song' to attract females. Rival males in close proximity to each other can alternately stridulate ('sing') in a 'rivals' duet'. Females sometimes respond through stridulation as they orientate towards the male. 'Courtship song' follows and this can be elaborate and quite different to the 'normal song'.

Identification

Adults of all Irish species of grasshoppers, groundhoppers and bush-crickets are readily identifiable without magnification. There are five species of grasshopper, two species of groundhopper and five species of bush-cricket.

Groundhoppers resemble small grasshoppers because of the short antennae and relatively short legs. The obvious difference is in the pronotum, which is greatly elongated and extends over the abdomen. Groundhoppers do not sing and do not have hearing organs.

Bush-crickets can easily be distinguished from grasshoppers by their extraordinarily long antennae. Other differences are less obvious, for example, bush-crickets stridulate by rubbing their wings together, while grasshoppers stridulate by rubbing their long hind legs against their wings. The Oak Bush-cricket can do neither, but taps out a song with the hind legs. Bush-crickets detect sound by means of small 'ears' on the front legs; in grasshoppers these are on the underside of the front of their abdomen. With practice, it is possible to identify grasshoppers and bush-crickets by their chirping songs.

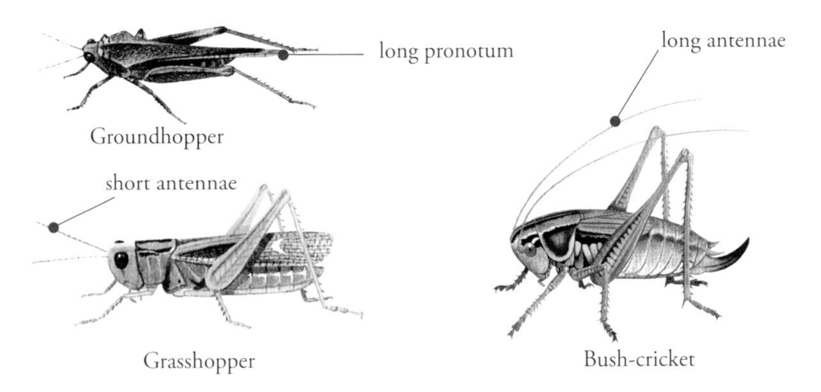

Groundhoppers

Common Groundhopper

Slender Groundhopper

Grasshoppers

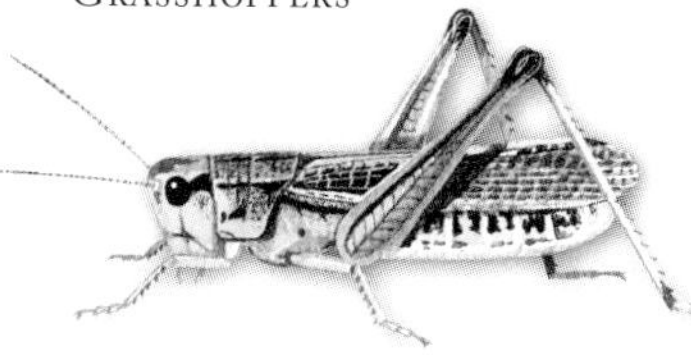

Large Marsh Grasshopper

Common Green Grasshopper

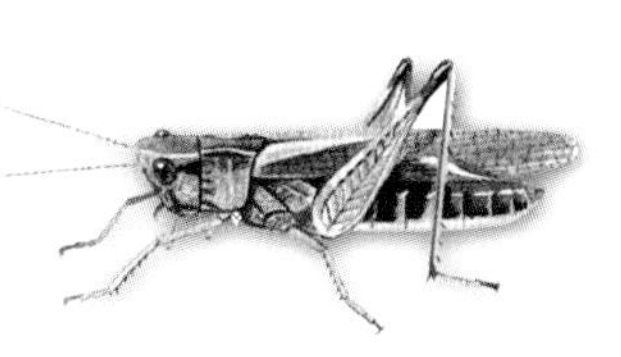

Field Grasshopper

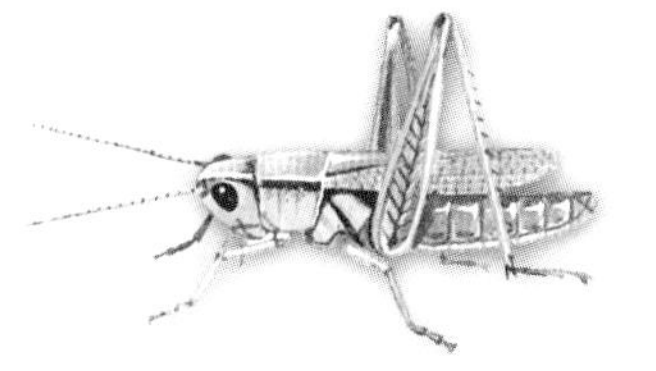

Lesser Marsh Grasshopper

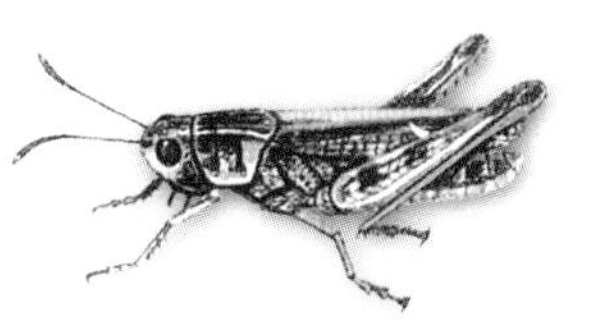

Mottled Grasshopper

Bush-crickets

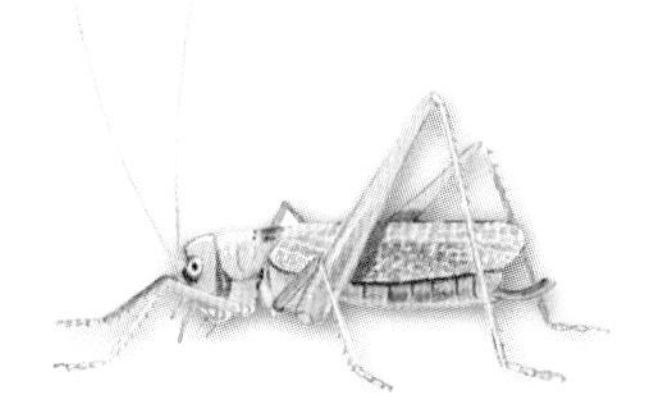

Oak Bush-cricket

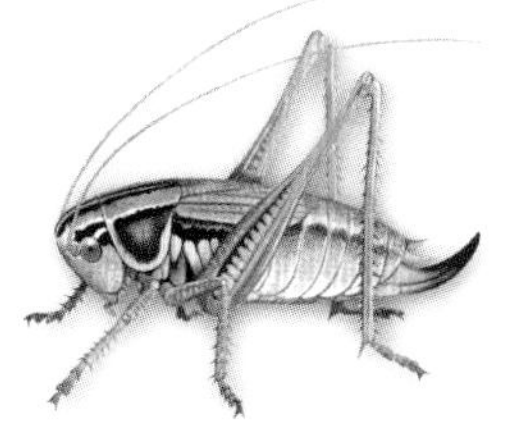

Roesel's Bush-cricket

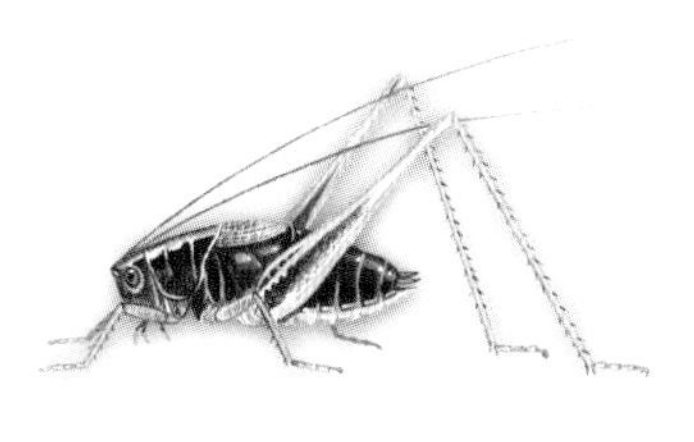

Dark Bush-cricket

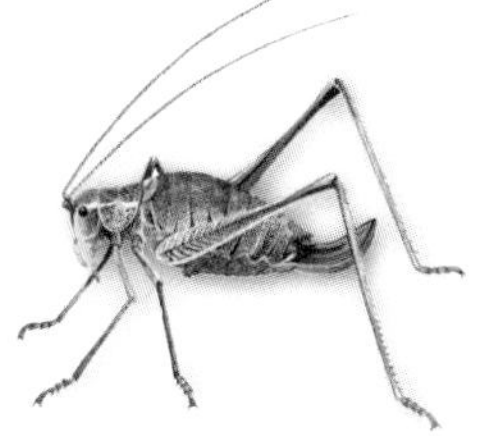

Speckled Bush-cricket

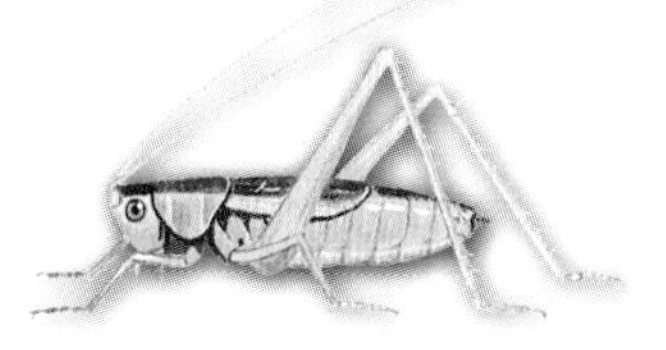

Short-winged Conehead

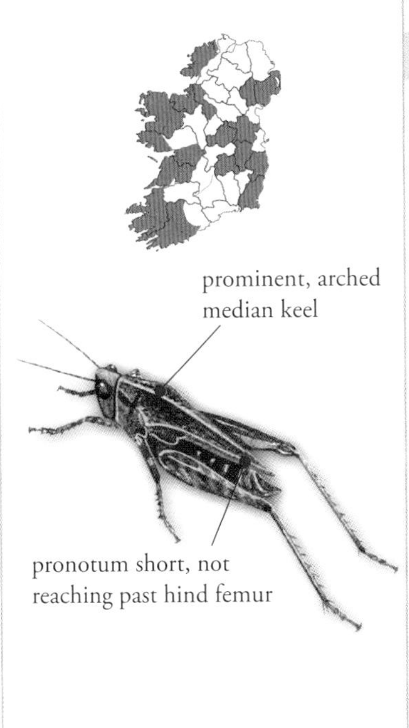

Common Groundhopper *Tetrix undulata*

Description: Length: males 8–9mm, females 9–11mm. Pronotum with a prominent median crest and not extending beyond the hind femur. Sandy brown to mottled grey or brown. A rare form has an elongated pronotum.
Season: Overwintering nymphs mature early in summer. Overwintering eggs mature later in summer. Nymphs hatching in later summer spend winter as nymphs and mature the following year.
Nature notes: Widespread and fairly common on bare ground or short vegetation with abundant moss including cutover bogs, dunes, quarries, heaths and marshes. They can swim and occur on lakeshores. Eggs are laid in clusters in soil or moss during spring and summer and hatch after a few weeks. Nymphs and adults feed on moss and algae.
Confusion: The Slender Groundhopper is similar but slimmer, with a less prominent keel and more elongated pronotum. Occasional specimens of the Slender Groundhopper have a short pronotum but still lack the prominent, arched keel.

Slender Groundhopper *Tetrix subulata*

Description: Length: males 9–12mm, females 11–14mm. Pronotum elongated and extending past the hind femur with a low, even crest on the midline. Occasionally with a shorter pronotum. Sandy brown to mottled dark grey and brown.
Season: Nymphs emerge between May and July but do not mature until the following spring. Eggs are laid from April until adults die off in July.
Nature notes: This species prefers calcareous soils and is most common across central Ireland where it may be found on bare mud with short vegetation in marshes. It also occurs in wet places on sand dunes. Eggs are laid in batches of 10–20 in moss or crevices in soil and they hatch a few weeks later. Nymphs have 5 or 6 instars and feed on moss and algae.
Confusion: The Common Groundhopper has a more prominent, raised crest. Occasional specimens have a very long pronotum but the crest is always more prominent.

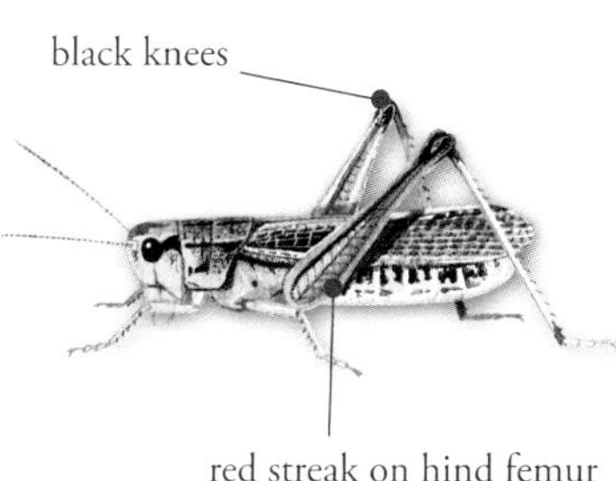

Large Marsh Grasshopper
Stethophyma grossum

DESCRIPTION: Length: males 22–29mm, females 29–36mm. Head, pronotum and abdomen usually vivid yellowish-green or olive-brown with darker mottled patches on the abdomen. There are usually red streaks underneath the hind femur. Hind legs usually with a yellow band adjacent to the dark knees.
SEASON: Nymphs emerge during May and June and mature by the end of July, surviving into October in mild weather.
NATURE NOTES: Only in very wet parts of blanket bogs and raised bogs but may be more widespread on intact raised bogs in central Ireland. Stridulation is distinctive amongst Irish species and is a series of clicks made by flicking the hind leg against the forewing. Eggs are laid in grass tussocks and the main food is grass.
CONFUSION: Distinctive due to large size and striking markings.

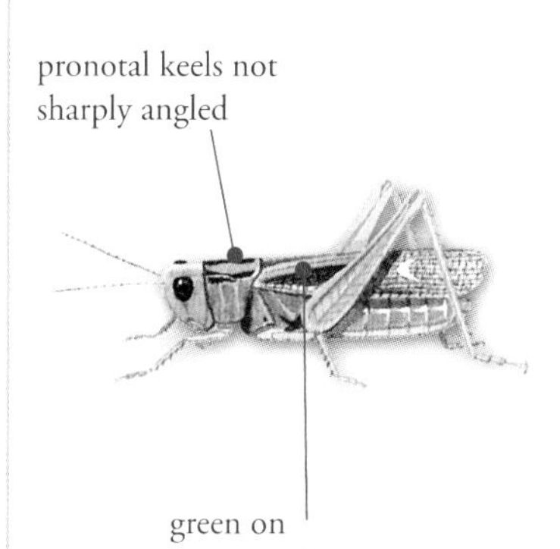

Common Green Grasshopper
Omocestus viridulus

DESCRIPTION: Length: males 14–18mm, females 17–23mm. Usually green on the upper surface of pronotum and wings, but with olive, grey, brown or sometimes purple on other parts of the body. Mature adults never have red on the abdomen. The pronotal keels are moderately curved but not sharply angled.
SEASON: Nymphs are active in April, earlier than other species, and become adults as early as the end of June. Adults rarely survive beyond October.
NATURE NOTES: Widespread and common in most parts of Ireland. Typically found amongst lush, coarse grasses particularly in meadows, bog margins and the edges of woods. Feeds on grass. The song begins quietly but gets louder. Eggs are laid in small batches in a pod inserted into soil where they lie dormant over the winter.
CONFUSION: The Field Grasshopper is similar but has more sharply angled pronotal keels and hairs on the thorax and it often has red on the abdomen.

Field Grasshopper *Chorthippus brunneus*

Description: Length: males 15–19mm, females 19–25mm. Hairy on underside of thorax. Body colouration varies from plain sandy brown, purple or even pinkish to mottled or striped. Males often have an orange-red tip to the abdomen. Pronotal keels sharply angled with a black wedge that does not reach the hind margin.

Season: Nymphs emerge in May and mature in June and July, moulting four or five times. Adults may live until November.

Nature notes: Widespread but scarce in the north. Found in dry, sunny places with short vegetation in grasslands, dunes and heaths. Males stridulate a short, repeated rasp, often calling and answering each other. Eggs are laid in batches in a pod inserted into soil.

Confusion: The Common Green Grasshopper has less sharply angled keels and much less hair on the underside of the thorax.

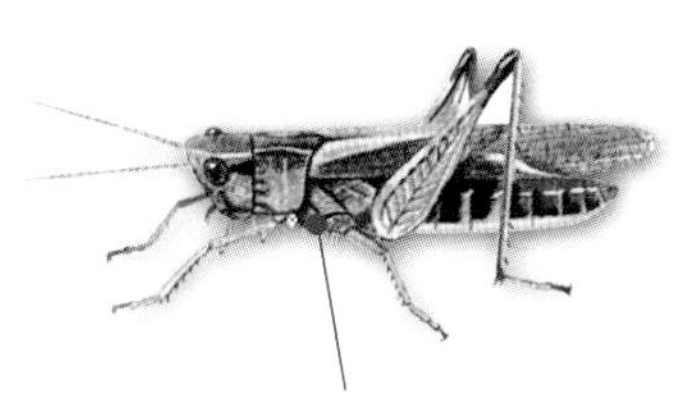

Lesser Marsh Grasshopper *Chorthippus albomarginatus*

Description: Length: males 14–17mm, females 17–21mm. Usually sandy-brown with green on the pronotum and wings but with no red on the abdomen. Pronotal keels are almost straight and parallel. Females may have a narrow white stripe on the forewings.

Season: Nymphs emerge in May and mature from early July. Adults may survive until mid-November in mild weather.

Nature notes: Only in the south and west near the coast in saltmarshes and damp places in sand dunes. Stridulation is a short rasp repeated 2–6 times. The calling song is a rapidly alternating whirr-click. Males and females fly readily. Eggs are laid in small batches in pods. Feeds on grasses.

Confusion: The Field Grasshopper is the most similar of the Irish species but it is hairier and has curved pronotal keels.

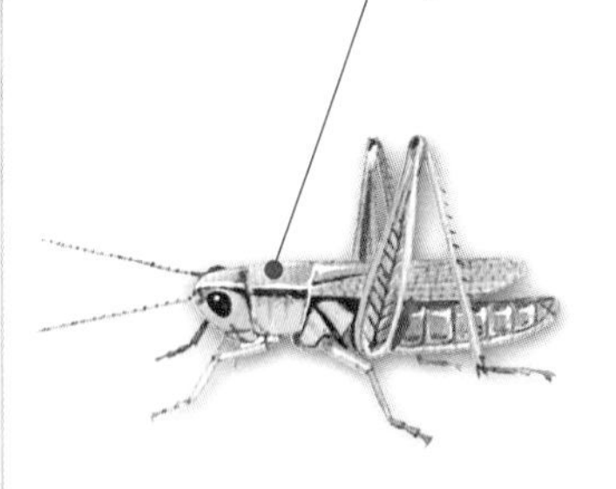

Mottled Grasshopper
Myrmeleotettix maculatus

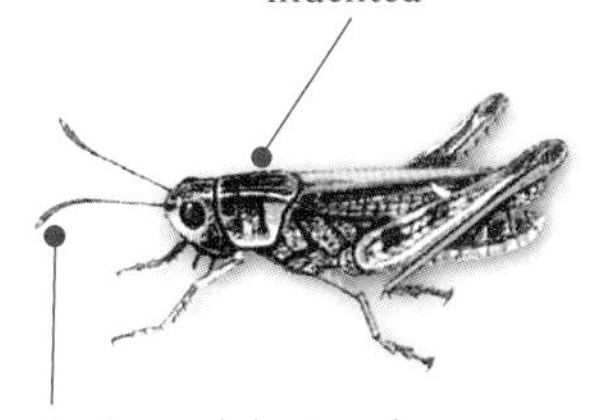

DESCRIPTION: Length: males 12–15mm, females 13–19mm. Wings are usually brownish and mottled with grey. Other parts can be mottled, green or red. Pronotal keels are white and are sharply indented and appear almost X-shaped. Antennae of males are bent outwards and have small knobs at the tip.
SEASON: Nymphs emerge between April and June and mature by mid-June, dying off by September.
NATURE NOTES: Widespread in well-drained and sparsely vegetated sites with some bare ground such as heaths, dunes, calcareous grassland and cutover bog. It spends most of the time on the ground. Males stridulate a series of buzzing chirps. Eggs are laid in batches in pods inserted into the ground and hatch the following spring.
CONFUSION: The sharply indented pronotal keels are distinctive and the characteristic club-tipped antennae easily identify males.

Oak Bush-cricket *Meconema thalassinum*

DESCRIPTION: Length: males 12–17mm, females 14–17mm. Pale green with thin brown stripe on the back. Wings reach the tip of the abdomen. Male cerci are in-curved. Females are slightly larger with a slightly up-curved ovipositor half as long as the body.
SEASON: Nymphs emerge around June and mature by late July and survive until November.
NATURE NOTES: Scarce in the west and southeast on oak and broadleaved trees in woods, hedges and scrub. Nocturnal and attracted to light. Mainly carnivorous, feeding on soft-bodied invertebrates. Females lay eggs in crevices on tree trunks. Most likely to be found by beating branches of trees and shrubs. Males lack a stridulatory organ so produce short rapid sounds by drumming with the hind feet.
CONFUSION: The Southern Oak Bush-cricket (*Meconema meridionale*) has been reported from urban gardens in Cork and Dublin since 2008. Adults are identifiable by the very short wings.

Roesel's Bush-cricket *Metrioptera roeselii*

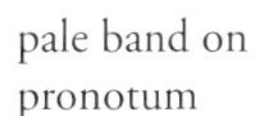

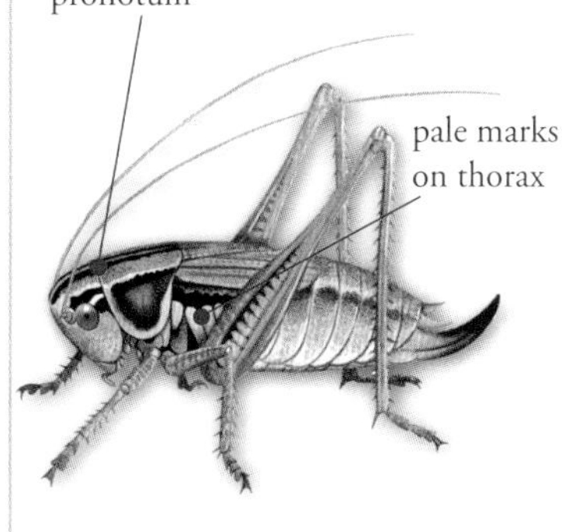

Description: Length: males 13–26mm, females 15–21mm. Brownish above, yellow underside, sometimes varying amounts of green. Cream band on the edge of the pronotum and yellow patches on the sides of the thorax and abdomen. Both sexes have short wings not reaching the end of the abdomen. Males have cerci with a small tooth on the inner side. The ovipositor is up-curved near the base.
Season: Nymphs emerge between May and June and mature by July. Adults may survive into October.
Nature notes: Only in coastal reed beds near the Blackwater estuary in east Cork and Waterford. Mainly vegetarian, feeding on grasses, herbs and flowers as well as soft-bodied invertebrates. Eggs are inserted into plant stems, often rushes. Males have a high-pitched, continuous chirp.
Confusion: The Dark Bush-cricket lacks the pale marks on the pronotum.

Dark Bush-cricket *Pholidoptera griseoaptera*

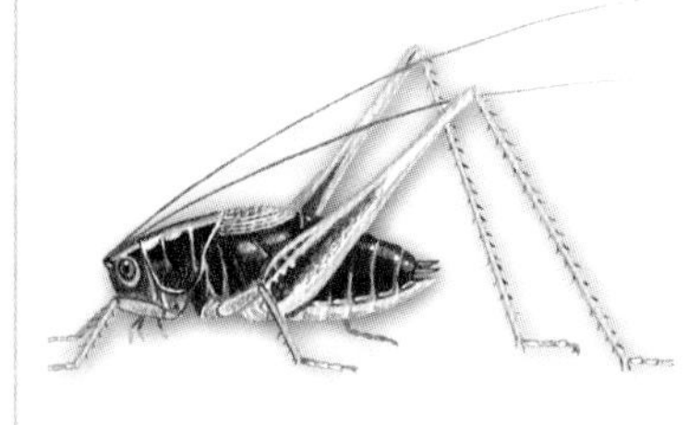

Description: Length: males 13–19mm, females 13–20mm. Generally dark brown above and with yellow-green underneath and always very short-winged. There is no keel on the pronotum and the forewings are very short, covering less than half the abdomen. Females have wings reduced to small lobes. The ovipositor is about half as long as the body, darkly coloured and upturned.
Season: Nymphs emerge around April and mature into adults by August. They are long-lived and survive into November in mild weather.
Nature notes: Only in mid-Cork, Clare and Waterford in thickets in scrub, rough grassland, hedges and woods. It feeds on plants, insects and spiders. Males sing a short chirp, repeated at regular intervals. Eggs are laid in bark or rotting wood.
Confusion: Roesel's Bush-cricket has cream-coloured marks on the pronotum and thorax.

Speckled Bush-cricket
Leptophyes punctatissima

DESCRIPTION: Length: males 8–16mm, females 11–18mm. Green with numerous small dark spots and a hunched appearance. Nymphs are densely speckled. There may be a brown stripe on the back. Males have short brown wings. Females have very short wings and a large, broad, up-curved ovipositor. Young nymphs resemble large aphids or green plant-feeding bugs
SEASON: Nymphs develop between May and July. Adults mature by August and may survive until November.
NATURE NOTES: Uncommon and only near south and east coasts from Cork to Dublin in bramble and scrub. It is solitary and feeds on various plants, including nettle, wood sage, honeysuckle and bramble. Males use their wings to make a high-pitched chirp. Females insert eggs into plant stems or crevices in bark.
CONFUSION: The Short-winged Conehead has a more pointed head and longer wings.

Short-winged Conehead
Conocephalus dorsalis

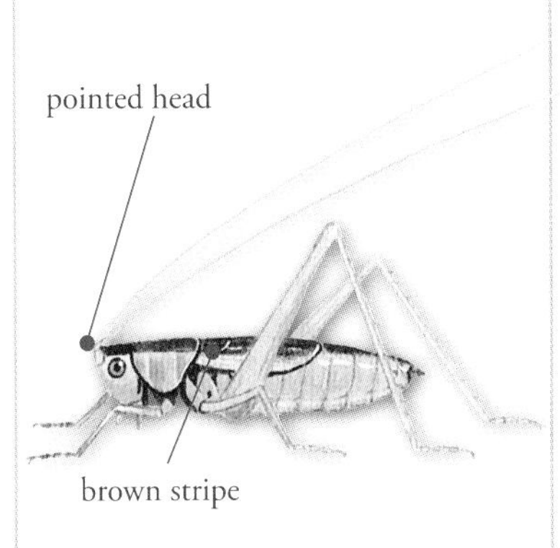

DESCRIPTION: Length: males 11–15mm, females 12–18mm. Adults and nymphs are green with a strong brown dorsal stripe. The head is pointed and the wings extend only halfway along the abdomen (although occasional, long-winged individuals occur). Specimens may also be all brown. The ovipositor is long, slender and slightly up-curved and brown.
SEASON: Nymphs emerge in May or June and mature by July and August. Adults may survive into October.
NATURE NOTES: First recorded in 1990 in County Cork and more recently in County Wexford. It lives in coastal marshes, saltmarshes and wet places inland with reeds and rushes. It conceals itself by lying fully extended against the stems of plants. Eggs are laid into stems of rushes, reeds and sedges.
CONFUSION: The Speckled Bush-cricket is finely speckled and more hunched.

Earwigs

Earwigs are among the most readily recognised insects and are best known for wielding fearsome-looking forceps. They are well known to gardeners because of the damage they inflict on seedlings and flowers. The vernacular English name probably derives from an old wives' tale, but they do seek out moist crevices in which to hide and may have, on occasion, entered ears. The name in German *öhrwurm* (ear-worm) and French *perce-orielle* (ear-piercer) also recall the terrifying prospect of earwigs taking up residence in ears.

Earwigs are one of the smaller insect orders with about 2,000 species described worldwide, only three of which occur in Ireland. They are active at night, feeding on a wide variety of insects and plants and they probably have some beneficial role in the garden as predators of aphids. The formidable-looking forceps or cerci on the tail end are used in defence and can give a painless nip. The forceps are usually bigger in males and are used for unfolding the wings, grabbing prey, mating and defence.

Earwigs have membranous wings folded in a complicated fan pattern underneath the short elytra. The order name Dermaptera translates as 'skin wings', although some species rarely use their flying ability. To unfold the hindwings, earwigs lift up their elytra and tease out the folded membrane with the aid of the forceps.

Life history

Earwigs do not have a pupal stage and moult five times before maturing into an adult. As far as is known, most earwig species exhibit maternal care, which is unusual amongst insects. Female earwigs construct brood chambers in crevices and guard their eggs and developing nymphs until the second moult.

Identification

With only three species of earwig in Ireland, they are easily identified. They are distinguished from other insects by the characteristic pincers or forceps. By far the most widespread and abundant species is the Common Earwig. The other two species, Lesne's Earwig and the Lesser Earwig, are quite uncommon.

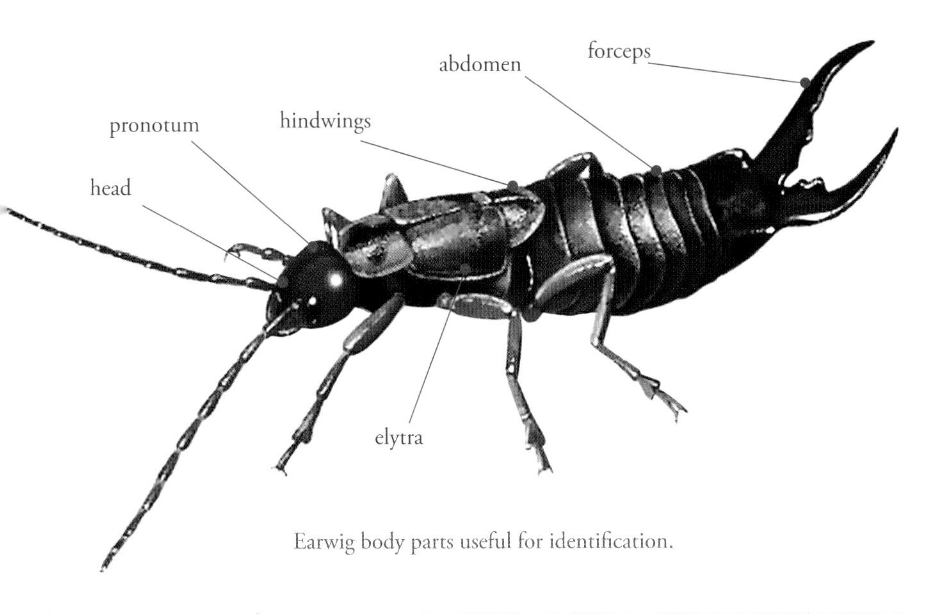

Earwig body parts useful for identification.

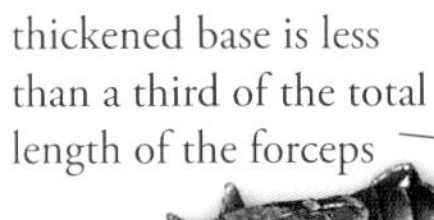

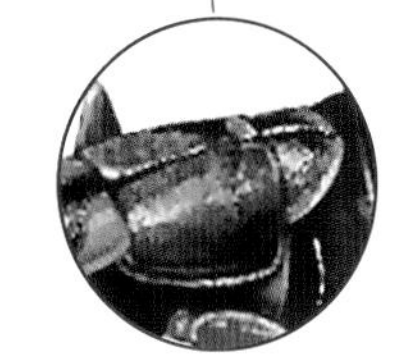

hindwings protrude from elytra

Common Earwig
Forficula auricularia Gailseach Choiteann

Description: Length: 10–15mm. The head is chestnut brown and the pronotum, elytra and abdomen are dark brown. The hindwings protrude from under the elytra as pale triangles. Males have long, curved forceps, which vary in length, particularly in isolated populations, such as on offshore islands. Females have shorter and straighter forceps. Nymphs are pale creamy white.

Season: Adults hibernate and emerge in spring to breed. Nests with eggs and nymphs can be found in spring and early summer. Nymphs mature into adults around July.

Nature notes: Common throughout Ireland in most habitats including grassland, woods, gardens and coastal habitats. It is nocturnal and a good climber and is often found on flowers. They are omnivorous and are known to eat petals and insect larvae. Eggs are laid in spring, in batches of 30–50, in a brood chamber. Nymphs resemble small, pale adults and are guarded by their mother during the early stages of development.

Confusion: Lesne's Earwig is similar but smaller and lacks protruding wings. The base of the forceps is relatively longer, about half the length of the forceps.

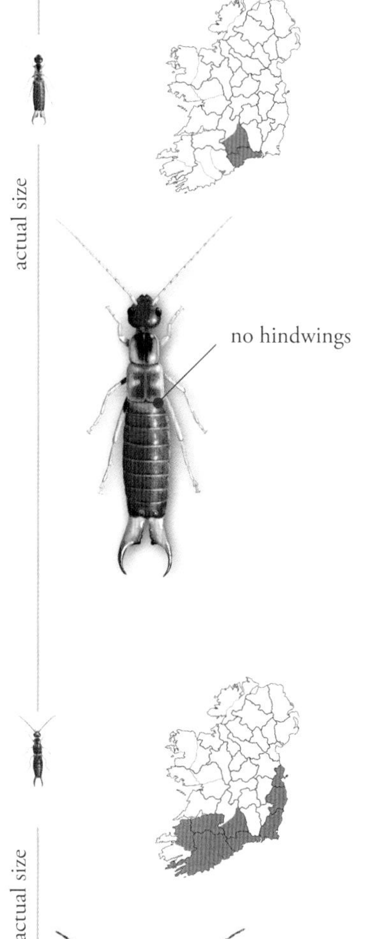

Lesne's Earwig *Forficula lesnei*

Description: Length: 6–7 mm. The head, pronotum and elytra are pale brown. Hindwings are absent. The abdomen is chestnut brown and barrel-shaped. The male forceps are strongly curved and have a thickened base approximately half the length of the forceps. Females have slightly curved forceps.
Season: Adults are mostly found in late summer but little is known about its life history.
Nature notes: Possibly a recent introduction to Ireland and, to date, known only from counties Waterford and Tipperary. Adults are nocturnal and may be found in leaf litter or plant stems in tall herbs and thickets in woods or rough ground. Like other earwigs, it is assumed to have maternal care. It is known only from the Atlantic fringe of western Europe.
Confusion: Smaller and paler than the Common Earwig and without hindwings.

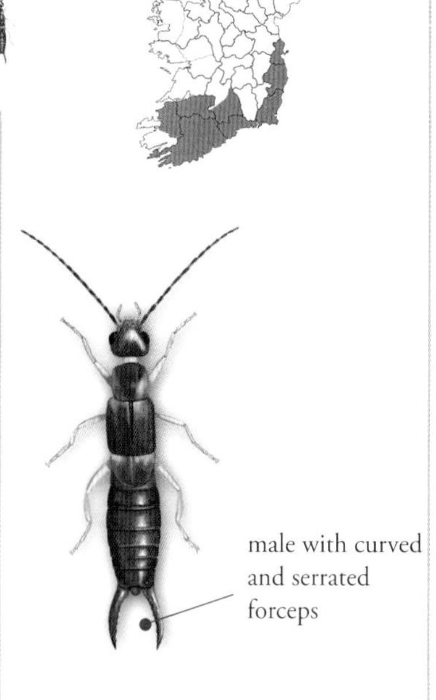

Lesser Earwig *Labia minor*

Description: Length: 4–6mm. Adults are chocolate or sandy brown. The hindwings protrude from under the elytra. The abdomen is chestnut brown and the legs are yellowish. Males have stout, gently curved forceps with serrations on the inner side. Females have straighter and more slender forceps.
Season: Breeds all year round and all life stages can occur together.
Nature notes: Apparently localised near the south coast but rarely recorded in Ireland and recently recorded only in southern counties. It lives in dung heaps or compost which provides a stable, warm, moist environment. Females guard and clean the eggs which hatch after 7–12 days. Nymphs remain with the mother for about a week and then disperse. Adults fly readily and are attracted to light.
Confusion: It is much smaller than the other two species.

Shield bugs

Shield bugs are 'true' bugs and are part of a large group of sucking insects that includes greenfly, plant hoppers, pond skaters and the dreaded bedbug. True bugs characteristically possess an elongated tube-like mouth called a rostrum and this organ is used to pierce and suck food. Many bugs are herbivorous and feed on plant juices, but some species are predators and a few are parasitic. There are over 770 species of true bug in Ireland and 17 of these are shield bugs.

Shield bugs are relatively large insects and some are indeed shield-shaped with broad shoulders, parallel-sided bodies and a tapered tail end. They are identifiable by their body shape and their partly membranous wing cases. Shield bugs could be confused with beetles but beetles never have membranous or partly membranous wing cases. Most Irish shield bugs feed on plant juices but some are predators, using their rostrum to stab soft-bodied animals such as caterpillars and beetle larvae.

Life history

Most of the Irish shield bugs overwinter as adults, hiding in sheltered places and re-emerging in spring to feed and breed. After mating, the female lays eggs on leaves, either singly or in clusters. The eggs hatch a week or so later and the shield bug nymphs begin to feed on plant juices or search for prey. The growing nymphs moult five times over a period of around six to ten weeks before becoming mature adults. Adults are most frequently seen in spring and late summer and nymphs are most abundant in midsummer.

Shield bugs are fascinating insects and they show some unusual behaviour. The female Parent Shield Bug carefully minds her eggs and young nymphs, protecting them from attack by predators and parasites. This is one of the few examples of insect parental care. The smallest Irish species, the tiny Scarab Shield Bug, has the ability to emit sound by rubbing abdominal segments together. Shield bugs can also defend themselves from predators by releasing a foul-smelling liquid from glands on their underside between the first and second pair of legs. For this reason they are sometimes called 'stink bugs'.

Identification

Adults of all the Irish species are readily identifiable without magnification. In some species, colours can vary with adults becoming less vibrant in winter but the features shown should lead to a positive identification. Another useful clue to identification is the host plant, however, they do stray from their hosts so it is wise to double-check the identification features shown. Most shield bugs live on vegetation, mainly trees and shrubs, and are most easily found by shaking branches over a sheet or sweeping through vegetation with a strong net.

Shield bug anatomy

The antennae are quite long and have four or five segments. The mouthparts form a long, needle-like tube called the rostrum, which is usually held close to the underside of the body. The first segment of the thorax is covered by a large broad plate called the pronotum. Behind the pronotum is a large, often triangular plate called the scutellum. The wing cases are fully developed when in the adult stage and are partly thickened and partly membranous. Nymphs are identifiable by the incompletely formed wing cases.

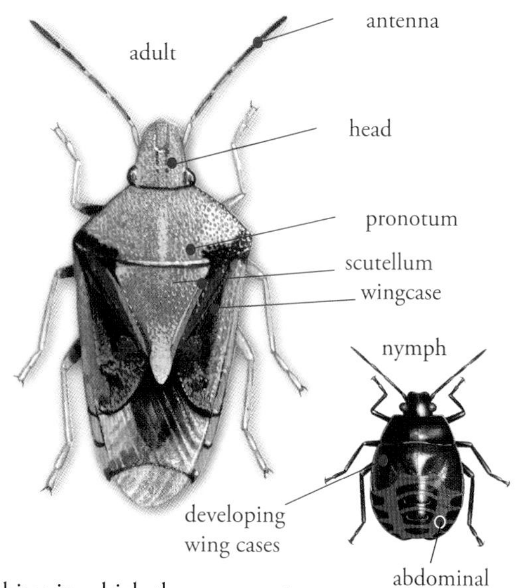

Shield bug habitats

Shieldbugs are grouped here by the habitat in which they are most likely to be found, although they can fly and may be found away from their usual habitat.

On deciduous trees and shrubs – nine species

Birch Shield Bug

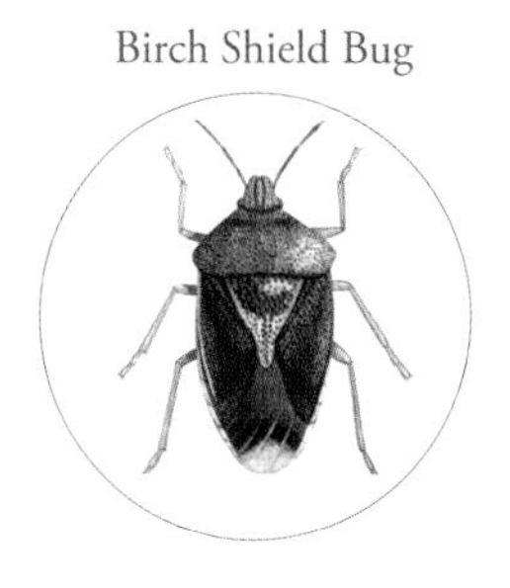

Bronze Shield Bug

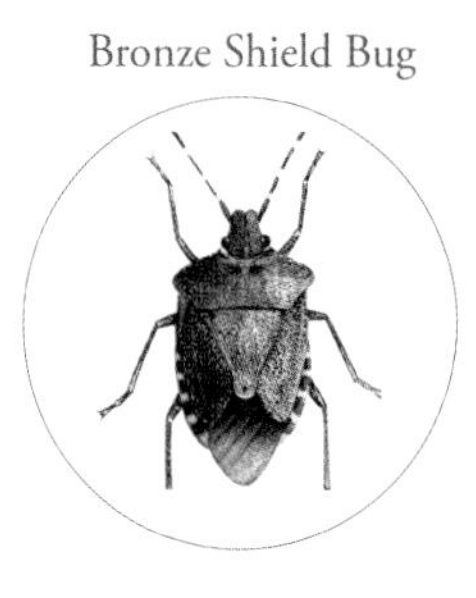

Forest Shield Bug

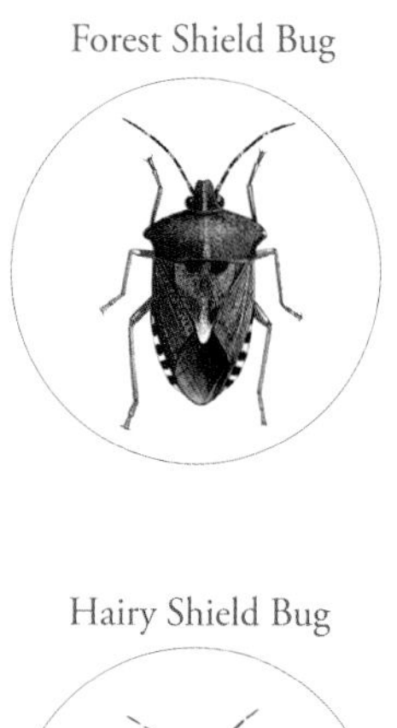

Gorse Shield Bug

Green Shield Bug

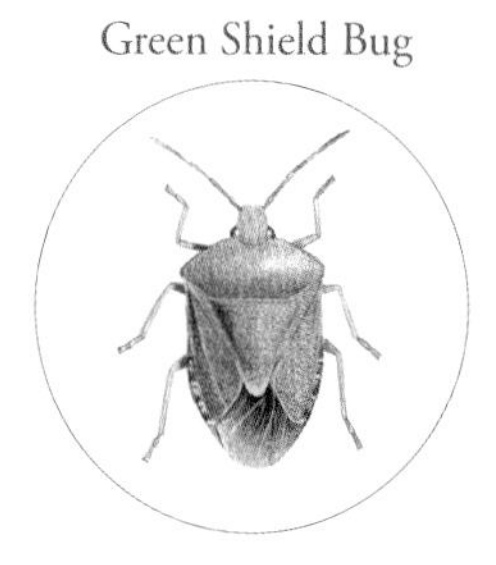

Hairy Shield Bug

Hawthorn Shield Bug

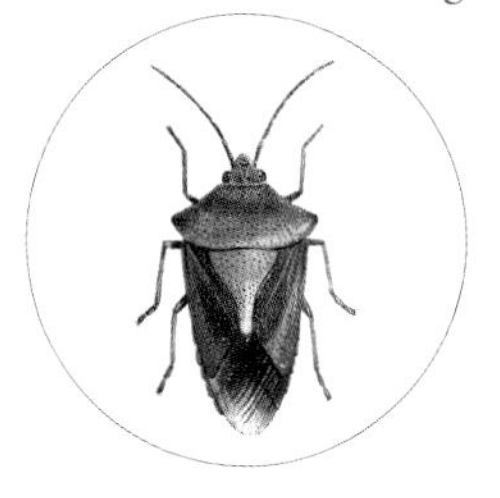

Heather Shield Bug

Parent Shield Bug

On dry sandy soil

Scarab Shield Bug

On coniferous trees – two species

Juniper Shield Bug

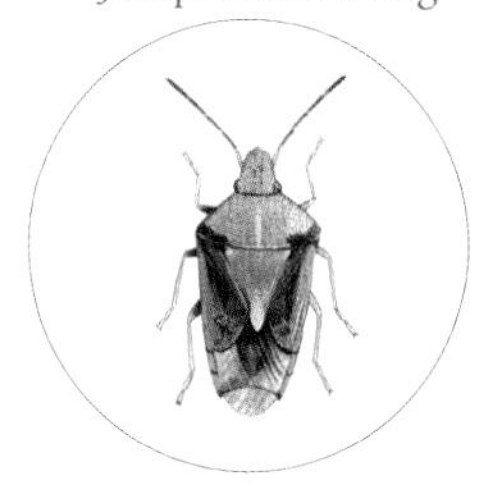

Western Conifer Seed Bug

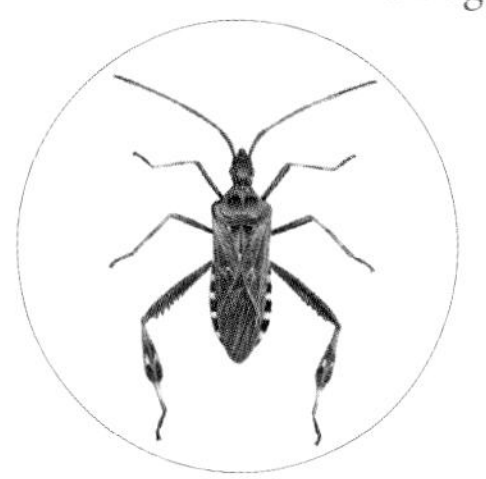

On herbaceous vegetation – five species

Blue Shield Bug

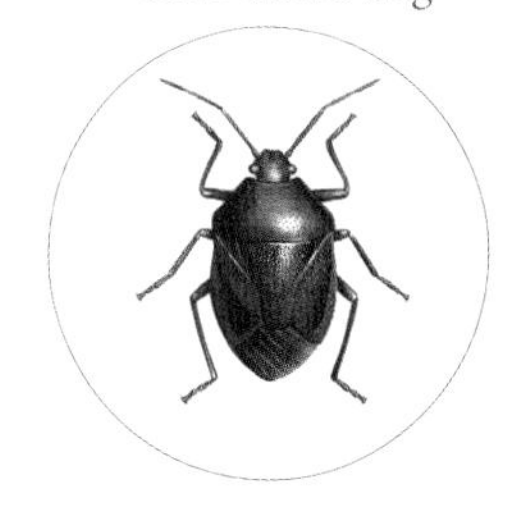

Dock Bug

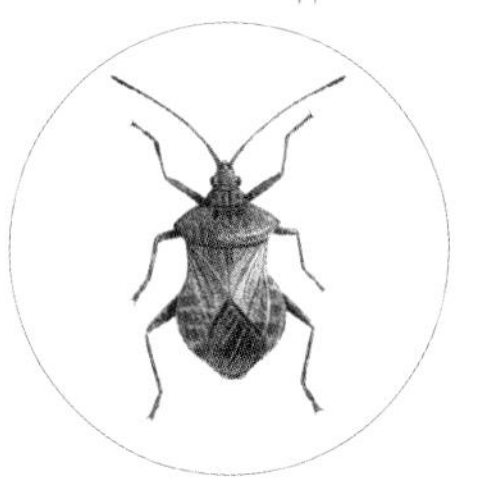

Forget-me-not Shield Bug

Spiked Shield Bug

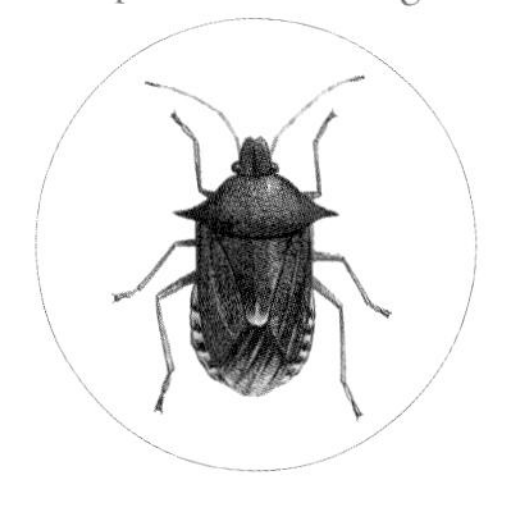

Tortoise Shield Bug

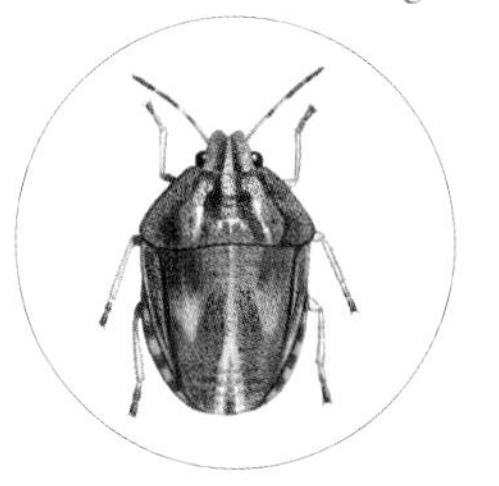

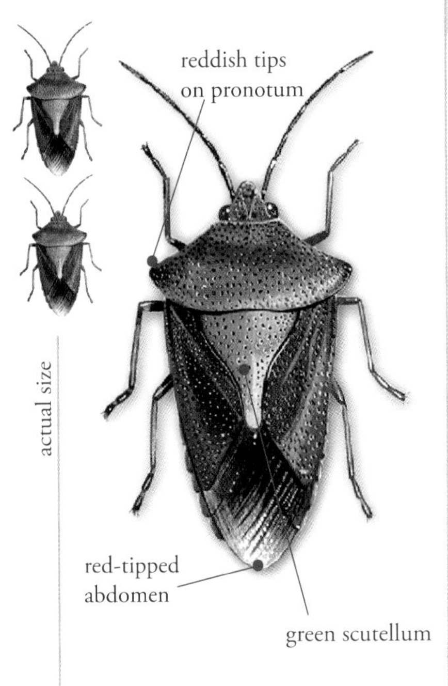

Hawthorn Shield Bug
Acanthosoma haemorrhoidale

DESCRIPTION: Length 13–15mm, bright green and reddish-brown shield bug. The scutellum is usually bright green and there are brown bands on the pronotum and wing cases. The noticeably protruding shoulders and abdomen are reddish.
SEASON: Adults may be found at any time of year but they become scarce in mid-summer. Eggs are laid from May to July and nymphs develop between May and October.
NATURE NOTES: Widespread and common all over Ireland occurring on trees in woods, hedgerows, parks and gardens. It is the most frequently seen shield bug and it feeds primarily on hawthorn, but also on other trees including oak, birch and hazel.
CONFUSION: The Birch Shield Bug has similar colouration but is smaller, with less pointed green-tipped shoulders.

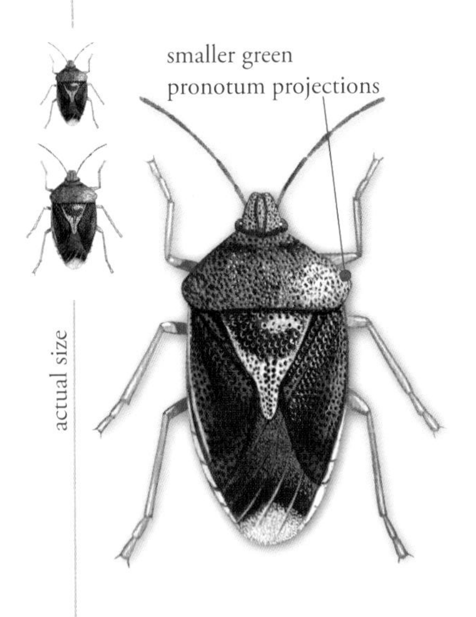

Birch Shield Bug
Elasmostethus interstinctus

DESCRIPTION: Length 8–12mm. The body is light green with dark punctures. The pronotum has a reddish-brown band on the rear edge and the shoulders only slightly protrude. The wing cases are marked with green, dark reddish-brown and black.
SEASON: Adults may be found at most times of the year except July and August. Overwintering adults emerge in spring to feed and breed. Nymphs develop on food plants during June, July and August mature into adults towards the end of summer.
NATURE NOTES: Widespread and common in hedgerows, bogs, gardens and parks. Adults and nymphs feed on juices of birch, hazel and aspen.
CONFUSION: The Hawthorn Shield Bug is larger with more protruding shoulders. The Juniper Shield Bug has more distinctive pinkish markings on the wing cases. The Parent Shield bug is less green.

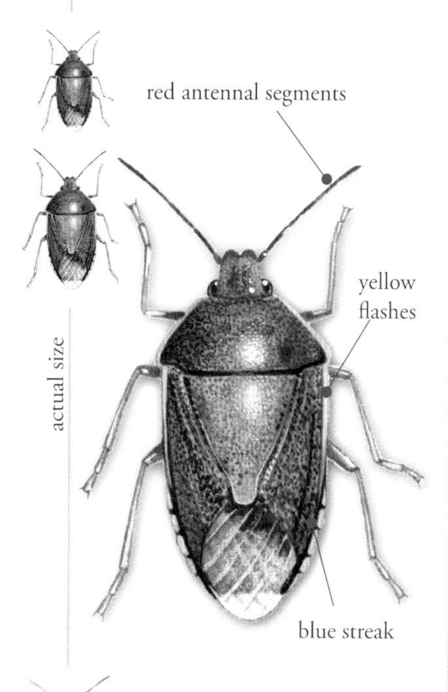

Gorse Shield Bug *Piezodorus lituratus*

DESCRIPTION: Length 10–13mm. In spring, the body is mainly yellow-green with grey-blue flashes on the outer edge of the wing cases. The antennae are red. In late summer adults develop reddish-purple bands across the pronotum and on the wing cases.
SEASON: Adults appear from August and survive through to the following July. Nymphs may be found between April and August.
NATURE NOTES: Widespread and fairly common in Ireland. This bug may be found on gorse and broom in heaths, bogs, parks and gardens. Nymphs and adults feed on leaves, flowers, seeds, and pods of gorse and broom and sometimes other members of the pea family.
CONFUSION: The Green Shield Bug is less coarsely punctured and has dark tips on the antennal segments. The late summer/winter colouration is similar to that of the Hairy Shield Bug but that species is hairy.

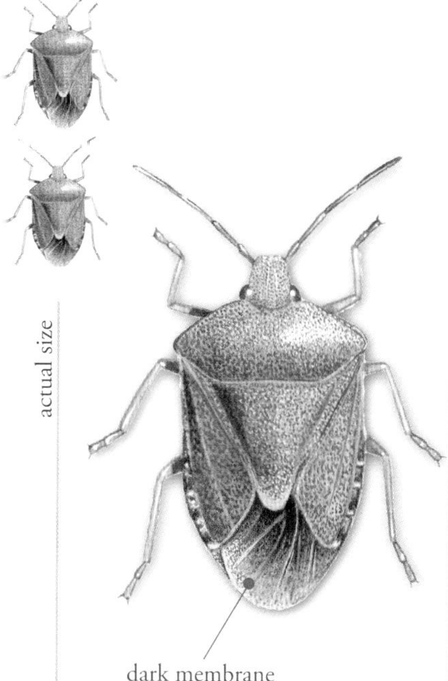

Green Shield Bug *Palomena prasina*

DESCRIPTION: Length 12–14mm. This large, green (turning brown in winter) shield bug is covered in very fine dark punctures. The membranous part of wing cases is dark and the antennal segments can be reddish with black tips.
SEASON: This species overwinters as an adult. Nymphs may be found between June and October. Adults appear from late August and may survive until the following July.
NATURE NOTES: Widespread in coastal counties in the east, south and west of Ireland but less common in the north and midlands. Found in hedgerows, parks and gardens. The nymphs feed on a variety of fruit and berry-bearing deciduous trees and shrubs, including hawthorn and bramble.
CONFUSION: The Gorse Shield Bug is similar but has a yellow streak and blue band along the side of the wing cases and pronotum.

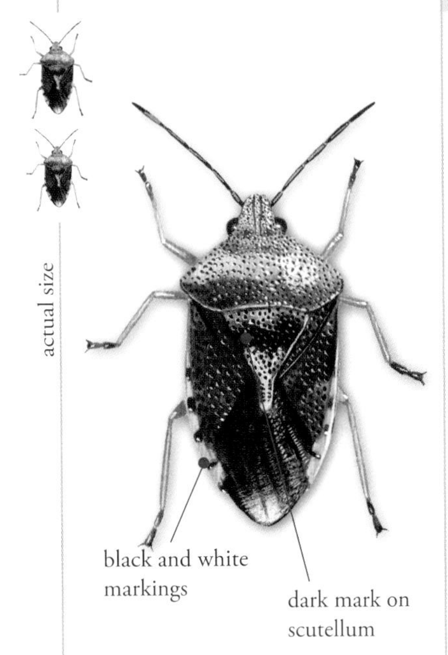

Parent Shield Bug *Elasmucha grisea*

DESCRIPTION: Length 7–9mm. Mottled grey on the body, with hints of brown, orange and black. There is a strong dark mark on the scutellum and black-and-white markings on the outer edge of the abdomen.
SEASON: Overwintering adults emerge from hibernation in spring. Nymphs develop on the host plant between June and August and adults may be found from August through to the following May.
NATURE NOTES: Widespread but not common in Ireland. Mates in spring and females guard the eggs and young nymphs. Often in higher branches on trees so possibly more common in Ireland than existing records indicate. Adults and nymphs feed on juices of birch and alder.
CONFUSION: Birch and Hawthorn Shield Bugs are somewhat similar but the Parent Shield Bug has more distinctive dark markings and less green.

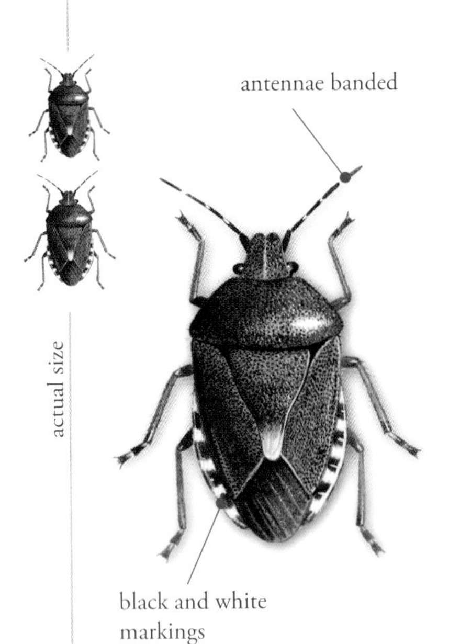

Hairy Shield Bug *Dolycoris baccarum*

DESCRIPTION: Length 11–12mm. Also known as the Sloe Shield Bug. In summer it is purplish-brown and green but in winter it becomes duller and more brown. The antennae and abdominal margins are banded with black and white.
SEASON: Adults appear in August and may survive through to the following July. Nymphs develop between June and September.
NATURE NOTES: Recorded from the southern half of Ireland and can be quite common. Associated mainly with tall vegetation in grassland, hedgerows, scrub and wood margins. Nymphs and adults feed on leaves, flowers and seeds of various shrubs and herbs especially blackthorn.
CONFUSION: The summer purplish and green colouration is similar to the Gorse Shield Bug's winter colouration. It is also superficially similar to the Parent Shield Bug but the Hairy Shield Bug is the only species covered with hairs.

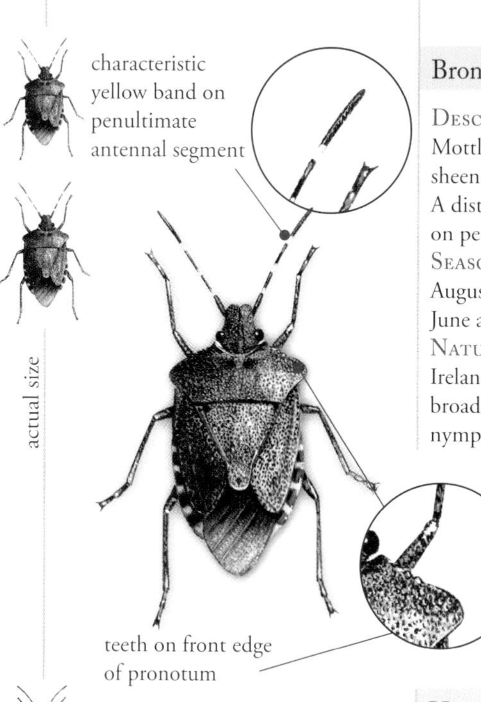

Bronze Shield Bug *Troilus luridus*

Description: Length 10–12mm. Mottled grey-brown with a metallic sheen on the body, legs and wing cases. A distinctive feature is the orange band on penultimate antennal segment.
Season: Adults may be found from August right through to July. Nymphs develop between June and September.
Nature notes: Widespread and fairly common in Ireland in woodlands, on the foliage of a variety of broadleaved and coniferous trees and shrubs. Young nymphs feed on plant juices. Older nymphs and adults are carnivorous and prey on the larvae of moths and beetles.
Confusion: The Forest Shield Bug has similar colouration except for the reddish legs and red-tipped scutellum.

Heather Shield Bug *Rhacognathus punctatus*

Description: Length 7–9mm. Dark metallic bronze all over with orange-red markings. There are red bands on the legs and an orange-red marking on the centreline of the pronotum.
Season: Adults appear in August and may survive until the following June and nymphs are found between June and August.
Nature notes: Uncommon in Ireland and only known from the north and southeast. It is found on heather and other shrubs including willows. This shield bug is predatory on beetle larvae, particularly the Heather Beetle (*Lochmaea suturalis*).
Confusion: The Bronze Shield Bug has similar overall colouration but is larger and does not have the distinctive bands on the legs.

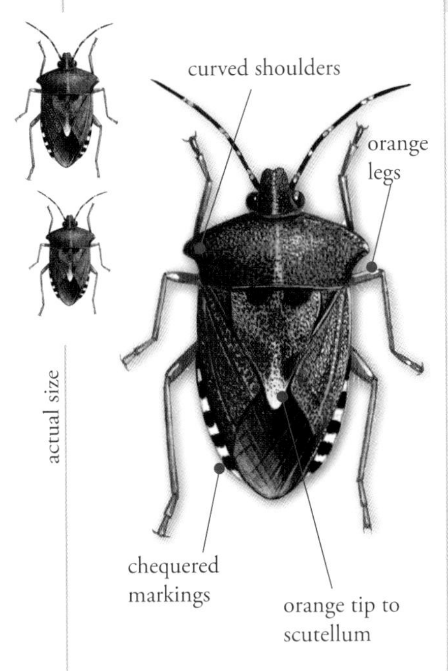

Forest Shield Bug *Pentatoma rufipes*

DESCRIPTION: Length 11–14mm. Pronotum and wing cases are uniformly glossy brown. Legs and tip of scutellum are orange. Curved shoulders protrude strongly and are blunt-tipped. Abdomen with chequered markings along each side.
SEASON: Nymphs overwinter, often high in trees, and may be seen all year round. Adults occur from July through to November.
NATURE NOTES: Widespread and fairly common in woodlands. It lives primarily on oak, but also on other trees in orchards, hedgerows and gardens. Nymphs feed on oak, alder and other broadleaf trees. Adults are partly predatory on caterpillars and other insects.

CONFUSION: Spiked Shield Bug and Bronze Shield Bug are similar. However, the Spiked Shield Bug has very pointy shoulders while the Bronze Shield Bug has two distinctive yellow bands on the penultimate antennal segment.

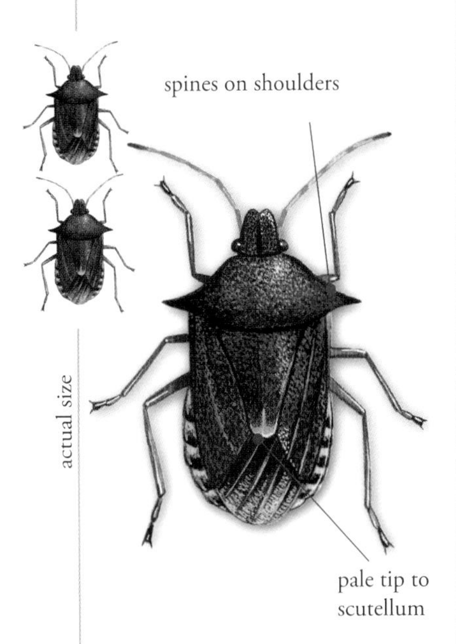

Spiked Shield Bug *Picromerus bidens*

DESCRIPTION: Length 12–13.5mm. Body brown with orange legs and antennae. It has very distinctive, sharply spiked shoulders.
SEASON: This species overwinters in the egg stage. Nymphs may be found between May and August. Adults appear from July through to November.
NATURE NOTES: Widespread from Kerry to Antrim but less common in the western half of Ireland. It is found in heaths, damp meadows, and bogs with long vegetation. Young nymphs feed on a range of plants. Older nymphs and adults are predatory, feeding on larvae of other insects.
CONFUSION: Forest Shield Bug and Bronze Shield Bug are similar in colour but the spiked shoulders are unique to the Spiked Shield Bug.

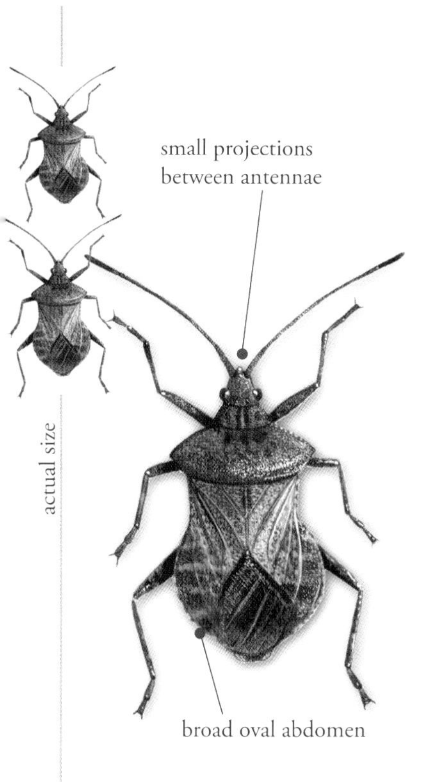

Dock Bug *Coreus marginatus*

Description: This mottled brown bug is relatively large at 13–15mm. Distinctive features are the broad, oval abdomen and two small projections between the antennae.
Season: Adult Dock Bugs may be found throughout the year. Nymphs develop from June to August.
Nature notes: In Ireland it is restricted to the southern counties. Found in hedgerows, waste ground or in dense vegetation where the food plants occur. Nymphs and adults feed on leaves and seeds of docks and related plants.
Confusion: This species is quite distinctive in Ireland because of the broad, oval abdomen.

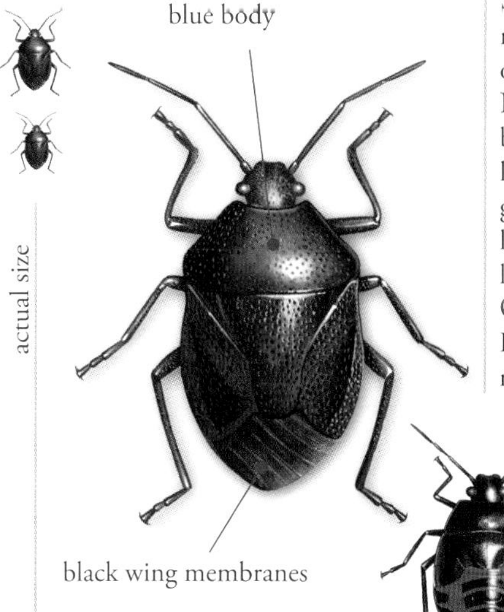

Blue Shield Bug *Zicrona caerulea*

Description: Length 5–7mm. Distinctively dark blue with a metallic sheen and dark wing case membranes.
Season: Nymphs develop between June and July and mature into adults by August. Adults overwinter and re-emerge in spring.
Nature notes: Widely distributed but not common in Ireland. Found on low vegetation in heaths, marshes, and grasslands. The Blue Shield Bug is most likely to turn up where its prey, leaf beetle larvae, is abundant.
Confusion: It is the only metallic-blue shield bug in Ireland but it superficially resembles some species of metallic-blue leaf beetle.

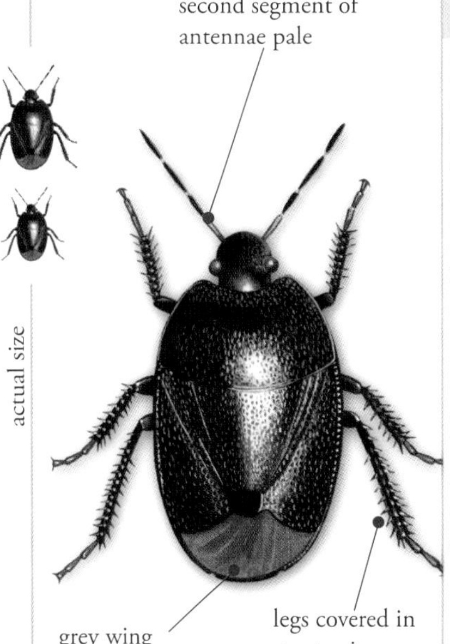

Forget-me-not Shield Bug *Sehirus luctuosus*

Description: Length 7–9mm. The body is black and shiny except for grey-brown wing membranes which overlap. The second antennal segment is paler than the others and the legs are covered in stout spines.
Season: Adults appear in late July and survive until the following spring while the nymphs develop during June, July and August.
Nature notes: Only known from old records from Kilkenny and Kildare but recently found near quarries in Co. Carlow. Associated with dry soils, cultivated fields and waste places. It spends most of its time on the soil surface but will climb plants to feed. As the common name suggests, this species feeds on leaves, seeds and flowers of several species of forget-me-not.
Confusion: The Blue Shield Bug is smaller and more bluish and the Scarab Shield Bug is much smaller.

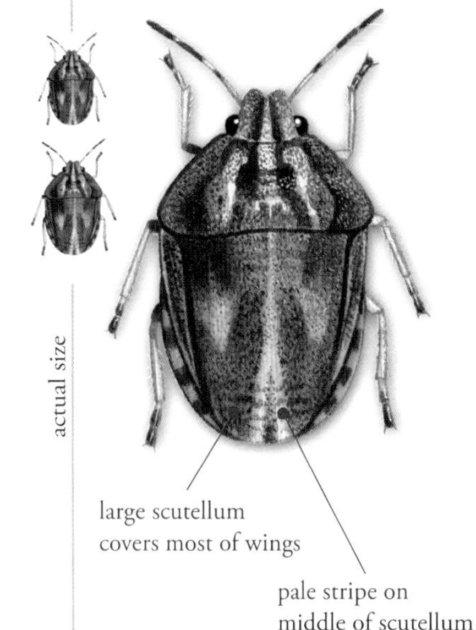

Tortoise Shield Bug *Eurygaster testudinaria*

Description: Length 9–11mm. This chunky shield bug has a greatly enlarged scutellum covering the most of the wing cases. Colouration varies from plain sandy brown to broadly streaked dark brown over a sandy-coloured background.
Season: Adults appear from late July until June the following year. Nymphs develop between May and August.
Nature notes: Widespread but uncommon across the southern half of Ireland, scarce in the north. In dry and damp tall grassland, woodland rides and hedge banks. Nymphs and adults feed on leaves, stems and flowers of various herbs, rushes and sedges.
Confusion: None, quite distinctive in Ireland.

actual size

large scutellum
covers most of the
abdomen

Scarab Shield Bug *Thyreocoris scarabaeoides*

Description: Length 3–4mm. This species is much smaller than other Irish shield bugs. It is completely metallic black with a bronze sheen. The scutellum is very large and covers most of the abdomen.
Season: Adults appear from late August and survive through the winter to re-emerge in spring and they may be found through to June the following year. Nymphs develop between June and August.
Nature notes: Last recorded in Ireland in the 1930s from Ballyteige, Co. Wexford, but it may be overlooked. It is associated with violets and is found in moss and leaf litter in dry, warm, sandy places such as coastal dunes. Nymphs and adults feed on leaves, stems, flowers and seeds of violets and pansies.
Confusion: The small size of this shield bug makes it very distinctive from other Irish species. Its large scutellum makes it look more like a beetle than a shield bug.

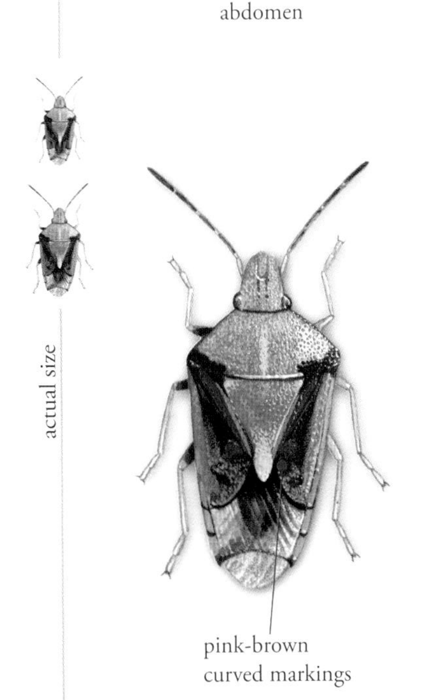

Juniper Shield Bug *Cyphostethus tristriatus*

Description: This is a mainly yellow-green, medium-sized bug measuring 9–10.5mm. The wing cases have distinctive, curved pink-brown markings.
Season: Adults may be found most of the year except August. Overwintering adults emerge in spring. Nymphs develop between June and September.
Nature Notes: First recorded in Ireland in Dublin in 1995 and since then near Portumna, Co. Galway. Found in woodlands and gardens on juniper and cypress trees. It may become more common because cypress trees are now very common in gardens throughout Ireland. Adults and nymphs feed on ripe berries and cones of the host trees.
Confusion: Birch Shield Bug is somewhat similar but lacks the distinctive pink markings.

Western Conifer Seed Bug
Leptoglossus occidentalis

DESCRIPTION: Measuring 16–20mm, this large shield bug is unmistakable in Ireland. The body is reddish brown with white W-shaped markings on each wing. The hind legs are broadly flared, unlike any other Irish species.
SEASON: Adults may be found in spring, summer and through the winter. Nymphs develop during summer mature into adults by late summer. Adults then seek sheltered places to overwinter.
NATURE NOTES: First seen in Ireland in 2009 in County Wexford. Lives on coniferous trees in plantations, parks and gardens. Nymphs and adults feed on sap from buds and seeds of conifers. It is native to western North America but has rapidly spread over much of Europe.
CONFUSION: None.

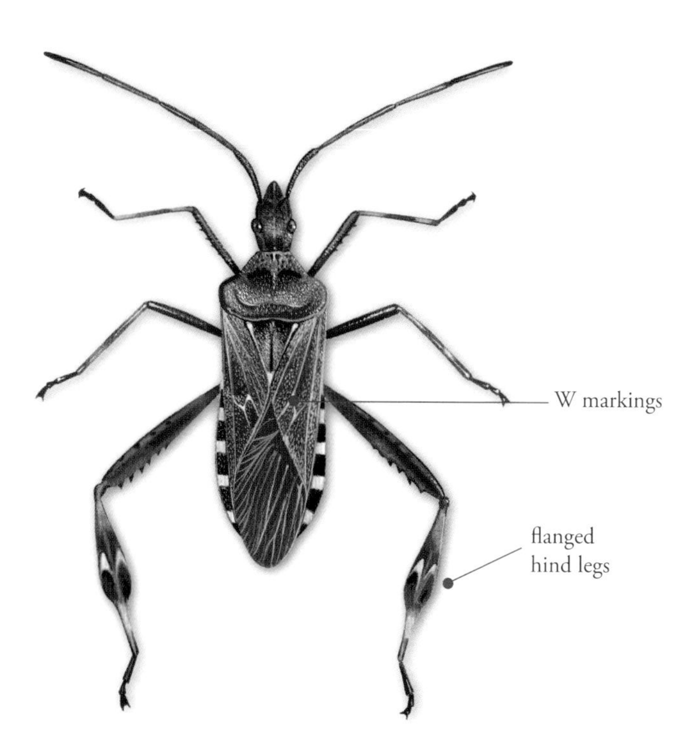

Glossary

Abdomen The hind section of an insect's body.
Antehumeral stripes Stripes on the upper surface of a dragonfly thorax.
Blanket bog Peatland formed in areas of high rainfall and covering expanses of undulating land. In Ireland, mainly in western and upland areas.
Bug An insect of the order Hemiptera.
Calcareous High in calcium, such as soil derived from limestone.
Caterpillar A larva of a butterfly or moth.
Cerci/cercus Paired appendages on the rear of insects, e.g. earwigs and crickets.
Chrysalis The pupa of a butterfly or moth.
Claspers Paired appendages on the rear of dragonflies.
Compound eye An eye composed of many individual light sensors.
Costa The leading edge of an insect's wing.
Crucifer A plant in the family Brassicaceae (formerly Cruciferae), including cabbages and relatives.
Ecosystem Service Resources and processes provided by nature.
Elytra Wingcases of beetles and earwigs (singular: elytron).
Exuvia The shed skin of an insect or other arthropod.
Femur Usually the largest and thickest segment of an insect leg.
Fen A peat forming wetland that receives nutrients from ground or surface water.
Frass Excrement of insect larvae.
Imago The final, fully developed stage of an insect.
Keel A long, straight ridge.
Labial palp A segmented sensory appendage beside an insect's mouthparts.
Larva The immature form of an insect that is markedly different from the adult.
Machair Grazed coastal grassland on sandy soil occurring mainly in west and northwest Ireland and often with high botanical and insect diversity.
Melanic With high levels of dark pigment.
Metamorphosis Changes in body form. It can be complete as in butterflies and beetles, etc, or incomplete as in grasshoppers and dragonflies.
Moult Shedding of the outer skin.
Nymph Immature stages of an insect that are quite similar to adult, i.e. insects with incomplete metamorphosis.
Order A group of related organisms with similar characteristics further divided into families, genera and species.
Ovipositor The egg-laying apparatus for females.
Parasitoid A parasite that eventually kills the host in/on which it feeds.
Pheromone A chemical released by animal that affects the behaviour of others.
Proleg A fleshy stumpy leg of a caterpillar.
Pronotum The dorsal surface or plate on the first thoracic segment.
Pterostigma A small coloured area near the wing tip of dragonflies.

Punctures Tiny pits or depressions on the surface of an insect.

Pupa The third stage of an insect with complete metamorphosis during which substantial changes in body form take place.

Raised bog A peat wetland that has developed in a former lake basin and is fed by rainwater.

Red List A list of organisms that are deemed to be under threat of extinction globally, regionally or nationally.

Rostrum A beak or snout, often the sucking, tube-like mouthparts of a true bug.

Sawfly Plant-eating insect related to wasps and bees, with caterpillar-like larvae.

Scutellum A portion of the dorsal surface of a thoracic segment. Sometimes it is visible as a small triangle near the front of the wings where they join.

Sex-brand A patch of pheromone-producing scales on the forewings of butterflies.

Stridulation Sound production by rubbing body parts together, e.g. grasshopper songs.

Thorax The middle section of an insect body, made up of three segments which each bear a pair of legs. Wings are borne on the second and third segment of the thorax.

Tibia The segment of the leg between the femur and the small foot segments

Umbellifer A plant in the family Apiaceae (e.g. cow parsley) with flower heads formed into level or rounded clusters that arise from a single point on the stem.

Vice-county A geographical division for the purposes of biological recording.

Bibliography

Brooks, S. and Lewington, R. 2002. *Field guide to the Dragonflies and Damselflies of Great Britain and Ireland.* 2nd Edition. British Wildlife Publishing. Rotherwick.

Edwards, M. and Jenner, M. 2005. *Field guide to the Bumblebees of Great Britain and Ireland.* Ocelli.

Evans, E. and Edmondson, R. 2005. *A photographic guide to the Shieldbugs and Squashbugs of the British Isles.* WGUK.

Majerus, M. and Kearns, P. 1989. Ladybirds. *Naturalists' Handbooks 10.* Richmond Publishing Co. Ltd. Slough.

Marshall, J.A. and Haes, E.C.M. 1988. *Grasshoppers and Allied Insects of Great Britain and Ireland.* Harley Books. Colchester.

Nelson, B. and Thompson, R. 2004 *The Natural History of Ireland's Dragonflies.* MAGNI Publication No. 013, National Museums and Galleries of Northern Ireland, Belfast.

Pinchen, B.J. 2006. *A pocket guide to the Grasshoppers, crickets and allied insects of Britain and Ireland.* Forficula Books. Lymington.

Pinchen, B.J. 2004. *A pocket guide to the Bumblebees of Britain and Ireland.* Forficula Books. Lymington.

Pinchen, B.J. 2004. *A pocket guide to the Shieldbugs and Leatherbugs of Britain and Ireland.* Forficula Books. Lymington.

Prŷs-Jones, O.E. and Corbet, S. 1987. *Bumblebees. Naturalists' Handbooks 6.* Richmond Publishing Co. Ltd. Slough.

Roy, H., Brown, P., Frost, R. and Poland, R. 2011. *Ladybirds (Coccinellidae) of Britain and Ireland.* FSC Publications. Telford.

Thomas, J. and Lewington, R. 2010. *The Butterflies of Britain and Ireland.* British Wildlife Publishing. Gillingham, Dorset.

Data on the Irish distribution of the insects covered in this book were taken mainly from the following online datasets:

Butterflies: http://www.butterflyireland.com
Ladybirds: http://www.habitas.org.uk/ladybirds/
Dragonfly Ireland: www.biodiversityireland.ie
Bees of Ireland: www.biodiversityireland.ie
Grasshoppers: www.orthoptera.org.uk
Earwigs: www.orthoptera.org.uk
Shield Bugs: Heteroptera of Ireland: www.biodiversityireland.ie

Internet resources

The Internet is a wonderful resource for entomologists, giving access to photographs, descriptions, maps, and scientific papers. However it changes regularly and the websites listed below may or may not exist by the time you get to them, but you can always use a search engine. We recommend the websites below as good starting points.

→ DragonflyIreland is an Ulster Museum website with photographs, descriptions, identification key and maps of Irish dragonflies and damselflies. See www.habitas.org.uk/dragonflyireland.

- → Ladybirds of Ireland is an Ulster Museum website with photographs, descriptions, and identification key for Irish ladybirds. See www.habitas.org.uk/ladybirds.
- → Ground Beetles of Ireland is another Ulster Museum website with photographs, descriptions and identification key for Irish ground beetles. See www.habitas.org.uk/groundbeetles.
- → Biodiversity Ireland has a range of sub-sites profiling various biological groups including bees and butterflies as well as links to biodiversity maps with up-to-date information on species' distributions. See www.biodiversityireland.ie.
- → Invertebrate Ireland is a portal site to information about the terrestrial and freshwater invertebrates of Ireland, putting checklists compiled by acknowledged experts in the Irish fauna online. See www.habitas.org.uk/invertebrateireland.
- → MothsIreland gives provisional distribution maps for 550 macro-moth species. See www.mothsireland.com.
- → Butterfly Ireland is a website of The Dublin Naturalists' Field Club with photographs, descriptions and distribution maps of Irish butterflies.
- → Butterflies and Moths of Northern Ireland is another Ulster Museum website with photographs, descriptions and distribution maps of all butterfly and some moth species of Northern Ireland.
- →Bugs, including shield bugs, are covered on the excellent website www.britishbugs.org.uk.

Further reading

Irish insects are a poorly studied group in general. However, there are a few guides that will aid anybody interested in learning more about the group. If you are interested in what is lurking in the corners of your house *Irish Indoor Insects: A popular guide* by James P. O'Connor and Patrick Ashe is an essential guide. Michael Chinery's *Insects of Britain and Western Europe* has an excellent key to the insect families with illustrations of a good number of species. With over 12,000 species, specialist identification books are needed to identify most insects to species level. Peter C. Barnard's *Identifying British Insects and Arachnids* is a good place to find the appropriate information.

The above are all concerned with insects in general. There is, however, a growing number of Irish books and identification aids covering just one group. The National Biodiversity Data Centre has produced an excellent suite of identification swatches including bumblebees, butterflies, shield bugs, and dragonflies while the National Museums of Northern Ireland have published a number of beautiful and comprehensive books on Irish insects, namely *The Butterflies and Moths of Northern Ireland* by Robert Thompson and Brian Nelson, *The Natural History of Ireland's Dragonflies* by Brian Nelson and Robert Thompson, and *The Ground Beetles of Northern Ireland* by Roy Anderson.

There are two excellent, independently published books on Irish butterflies: *Ireland's Butterflies: A Review* by David Nash, Trevor Boyd and Deirdre Hardiman and *Discovering Irish Butterflies and their Habitats* by Jesmond Harding. They can be obtained by contacting the Dublin Naturalists' Field Club and Butterfly Conservation Ireland, respectively.

Index of scientific and common names

Pearl

God uses our pain, suffering & obedience
to heal us in our darkest hour

JAMES V. DANIELS

Trilogy Christian Publishers
A Wholly Owned Subsidiary of Trinity Broadcasting Network
2442 Michelle Drive
Tustin, CA 92780

For information, address Trilogy Christian Publishing
Rights Department, 2442 Michelle Drive, Tustin, Ca 92780.
Trilogy Christian Publishing/ TBN and colophon are trademarks of Trinity Broadcasting Network.
For information about special discounts for bulk purchases, please contact Trilogy Christian Publishing.
Manufactured in the United States of America

10 9 8 7 6 5 4 3 2 1
Library of Congress Cataloging-in-Publication Data is available.
B-ISBN# 979-8-88738-723-9
E-ISBN# 979-8-88738-724-6

TABLE OF CONTENTS

INTRODUCTION

This is a message for all who have lost a loved one, whether recently or in the past, and are still suffering and in pain; where that pain is so unbearable you want to die yourself. You feel like giving up on life because there are no answers to your questions, such as "Why did this happen?" "She was so young and a good person." "He didn't deserve to die; he was kind to everyone." "Why didn't it happen to me instead of him or her?" These are some of the things you ask yourself. Then you turn to God and ask the big question: "Why, God? Why did this have to happen?" in your shock and disbelief. Your life will never be the same again because of the loss of your loved one.

I would like to start off by saying if you live long enough you will experience the loss of a loved one--father, mother, spouse, sibling, close relative, close friend, people you know of and encounter daily. Like your neighbors, people at your job, the cashier at the supermarket, the barber, famous people you hear about on the news, and the list goes on. But still by knowing this, it doesn't take away the hurt, the pain and the emptiness you feel inside from the loss of your loved one. Knowing that death is part of life doesn't make it one bit easier.

I would like to share a story about a Pearl. She was my wife, and she went home to be with the Lord. I'm a military retiree; I met my wife while stationed in Hawaii. She was a schoolteacher from San Diego, California, on vacation. We knew each other for only nine months and were together for just three of those months before we got married. My wife was a beautiful person, with her warm smile, her kindness and her loving heart. She was an amazing wife to me and our three children; she loved her family. Pearl was my best friend for nearly 38 years, with a zeal for life and a caring soul for all people. Always placing the needs of others before her own, she touched the lives and hearts of all she encountered. Two of the things I admired about her were her positive attitude and her strong drive. She always found the good in everything.

Pearl was an encourager, a motivator, and a very good friend to many people. She loved children and always brought out the best in people, because she loved them. Even if someone did something bad to her, she did not hold a grudge against them or seek revenge. My wife just left them alone and loved them from afar. She would say, "If you know that dog bites do not go back into that yard again."

She loved stores, shopping, and spending most of her money on people. Almost every other month Pearl would buy at least ten of the same items. We would have discussions about why we needed ten of whatever it was. I would always say to her during our discussions, "We are going to be homeless if you keep this up!" My wife would reply with a smile on her face, "One's for the people at the gas station that pumps gas in the car. One's for our mailman and his wife, another one's for the people at the grocery store who help put the groceries in the trunk of my car."

Then I would say, "You're giving all our money away."

"You'd probably give me away too if I let you." She would just smile and laugh and then she would say, "No, I'm going to keep you," and lastly, "It's only money; you can't take it with you." And that would be the end of our discussion.

I would also hear that still small voice at times from the Holy Spirit say, "Remember, she did tell you before you guys got married that she liked to shop." I would smile on the inside after hearing that voice.

She loved preparing food for her family. She loved Las Vegas, the food, and activities. She liked reading books by Joyce Meyer and T.D. Jakes. She also loved reading romance novels.

She was a member of the local church in our community for the past sixteen years. She loved the Lord. Her favorite Bible scripture was "And we know that all things work together for good to those who love God, to those who are the called according to His purpose. " Romans 8:28 (NKJV)

I thank God for the time I had with my wife.

Though she is not here with me physically, I know she is still with me in spirit and alive today. The Bible says in 2 Corinthians 5:8, "To be absent from the body is to be present with the Lord." I know that I will see my wife again one day in heaven and that she is waiting for me.

For those that have lost a loved one, who are looking for answers to your questions and are also looking for comfort, looking for strength, looking for peace, and are crying out for help, I will tell you: look to the Lord first. Psalms 28:7 says, "The Lord is our strength," and Psalms 121:2 says, "Our help comes from the Lord." God is still nurturing me, and I am getting stronger every day because I continue to trust in God and what His Word says. He strengthens and helps those who seek Him.

I hope by sharing my life experiences in *Pearl*, I will help answer some of the questions people may have about the loss of their loved ones, or a trial or tribulation you are facing, or anger you may be experiencing. Whatever challenges you are up against, I pray that God heals the pain and suffering you are going through. I pray that

He gives you strength, peace, comfort, and love, because God is love. I pray that He lifts you up and restores you, better than before. And I give God all the praise, the glory, and the honor for what I'm about to share with the world. In Jesus' Mighty Name I pray, amen.

MONDAY

It was on a Monday that my life would change forever. Seven days before that Monday, Pearl was hospitalized. She was very ill and the doctors thought she might have suffered a heart attack to go along with her other illnesses. The cardiologist's report indicated that her heart showed signs of a previous heart attack. They asked us if my wife had ever experienced heart problems or had a heart attack in the past. My children and I were baffled; as far as we knew my wife always had a healthy heart. The doctor gave my wife little chance of pulling through.

I had already prayed to God for my wife's healing and believed that she was healed. I was putting my faith in God, believing for the unseen. 2 Corinthians 5:7 says, "For we walk by faith, not by sight," and 2 Corinthians 4:18 says, "While we do not look at the things which are seen, but at the things which are not seen. For the things which are seen are temporary, but the things which are not seen are eternal." I was standing on the Word of God, period. I was believing what God's Word says and not what the doctor's report says, because God has the final say so. Like I already said, I believed that she was healed. My pastor, the church, and friends from all over the world prayed for Pearl's healing.

I was praying for my wife. I was meditating on healing scriptures day and night. Psalms 107:20-- "He sent His word and healed them from their destruction." Isaiah 53:4-5 and 1 Peter 2:24, "By His stripes we were healed." "Faith cometh by hearing and hearing by the word of God." (Romans 10:17) I was praying in tongues. I was doing everything that I knew to do that lined up with the Word of God.

Finally, we got some good news that my wife's condition was turning for the better. The medicines the doctors were giving her were finally starting to work. But the doctors said she still was not out of the woods yet. They were worried about her kidneys. Still, we continued praying over my wife, trusting and believing that God had already healed her. I received more good news the next day from a representative that worked with the hospital. This person asked about providing a nurse for my wife when she was discharged from the hospital. The representative asked if we had a downstairs bedroom. She said that my wife was going to need a nurse around the clock until she fully recovered. I was like, "Thank you, Jesus, that's another good sign that my wife is coming home soon."

Monday, a day or two after receiving the positive news of my wife's condition and the blessings of prayer from the pastor, friends and everyone, I was on the freeway driving to my oldest son's place, with my daughter. I received a call from the nurse at the hospital to come now, that she was scared for my wife, and she informed me that my wife's condition was starting to decline. Immediately I hung up, explained the situation to my daughter, and handed her my cell phone to call her siblings, while I looked to exit the freeway and turn around. On our way to the hospital, I was saying to myself, *She was doing so well... how could this be happening?* Then I told myself, *I'm standing on the word of God. She is already healed. No fear, no doubt; God has the final authority.* At that moment my daughter asked me if I was okay because I wasn't really focused on the road and was driving a little faster than usual. I told her I was okay and regained my composure. When we arrived at the hospital, my middle son was already there, and had arranged for us to go up to my wife's room. Surprisingly my oldest son, whom my daughter and I were on the way to visit, was already there at the hospital too. I said to myself at that moment, *I don't know how fast he drove to get here.* But thank God, he made it safely.

As we all made it up to my wife's room, the doctor and nurses were trying to stabilize my wife's blood pressure. All we could do was watch through the glass window of her room as they worked. My wife was aware that we were all there. The hospital had already arranged

Face Time cameras so that my wife could see and communicate with us and set up chairs for us to be comfortable. Myself, kids, and others took turns talking to my wife as the medical team continued to try and stabilize her blood pressure. She had a breathing tube, but she still communicated with us by blinking her eyes.

Finally, they were able to stabilize my wife. I was thanking God, giving Him all the praise and glory. We continued communicating with my wife until we were no longer allowed to be in the ICU. But we were allowed to wait in the ICU waiting room. While we were there the doctor came in and gave us a negative report and his opinion. I told him that I wanted him to do everything possible to help my wife. I also communicated to him that I was trusting God for my wife's healing. The doctor then said he was also a believer. When the doctor said he was a believer, I thought to myself, *How can a believer speak without hope?* I then looked the doctor straight in the eyes and said to him, "My wife is a fighter and a believer too, and I asked the doctor again to do everything medically possible to keep my wife alive. I also told him that he was to treat her like she was the President of the United States. Lastly, the bottom line: we're trusting in God.

The doctor left, then my daughter had a moment, saying out loud, "This is unfair" or words to that effect. I felt the same way my daughter felt at that moment. How could this be happening to a person who had given her whole life to helping people? It just wasn't fair. I kept how I felt on the inside. I had to remain strong in faith and try to set that example for my kids even though they were all adults. We had to continue to believe and have no unbelief in that waiting room.

We finally left the hospital and said our goodbyes to my wife and let her know that we would be back the next day early to see her. The nurse verified my phone number again and my kids' phone numbers. It was a long day at the hospital, but I was thankful that my wife was stabilized and doing better. It was very late when I got home; my daughter and my wife's brother stayed at the house with me. In about 30-40 minutes I would receive the worst news of my life. My cell phone rang. It wasn't the hospital; it was my middle son. He said, "Dad, Mom just passed away." At that moment I became angry, I wanted to go crazy, and I thought I was going crazy. The next thought was anger towards the doctor. *You call my son first to tell him that my wife passed instead of me first. You had my number and had been calling me first about everything until now when she passes.* The devil had my thoughts momentarily. I regained my spiritual composure; the Holy Spirit grabbed a hold of me. I then said to my son, "How did everything happen?" He explained his mom's passing to me. Then I asked him why the hospital didn't call me first. He said that they told him they couldn't get ahold of me first, that's why they called him. I'm thinking to myself again, *That's a likely story, because when I checked my phone there were no calls except the one I got earlier that day from the hospital.* The Holy Spirit said to me, "Let it go, James." I let those thoughts go from the devil. I then told my son thank you, we spoke for a couple of more minutes, the call ended. I had to tell my daughter and brother-in-law who were in the other rooms of the house and call my oldest son. I got myself and my thoughts together. Then I said to God, "I'm still trusting You and I'm not angry at You." Little did I know at that moment after I said that to God, weeks later I would get angry at God as I grieved for my wife.

Thinking back to that Monday, it was just like she was holding on to say goodbye to her family before she went home to be with the Lord. Knowing my wife, she probably saw glory and chose heaven over being in that sick body here on earth. My wife always loved venturing out, meeting people, seeing and doing new things and sharing them with her family. I can hear her right now on that day she left us talking to Jesus, saying, "Come on, Jesus, I want to see heaven. I'll see them later when they come; they'll be all right." That would be my wife 24/7, 365. I know when my time comes and I meet up with her in heaven, the first thing she'll say to me is, "What took you so long?"

UNBELIEF AND GUILT

It was unbelievable. I had just lost my wife, my best friend, the love of my life and the mother of my children. I thought, *God, how could this be happening?* It was like being in a dream; everything that had just happened seemed unreal. I couldn't accept the fact that my wife had gone home to be with the Lord. She was no longer with us in the physical body here on earth. The thing that hurt the most: there was nothing I could do about it. I was a Christian, a follower and believer of God. I study and read God's Word daily, I attend church and serve in the ministry, I attend Men's Bible Study Thursday nights and I minister to people that I meet daily because I know that I could be the only Jesus that some of them may know. I'm not trying to put myself on a pedestal or sound "holier than thou," but I believed in God for her healing, and we didn't get the healing. He could have kept her here with us.

After my family didn't get the healing, it led me to question initially, was it my unbelief, lack of faith, or lack of knowledge that she was no longer here, was that the reason we didn't get the healing? I would learn later the guilt I was experiencing was placed in my thought pattern by the one and only Satan. He has only one weapon: to cast doubt and attack your mind. Through your mind is the only way the devil can get to you. There is an eye gate and ear gate which are connected to your mind also, what you hear and see. Proverbs 4:20-21 refers to this.

In 1 Peter 5:8, it tells us to stay alert and watchful because the devil, who is our adversary, walks around like a roaring lion, seeking someone to devour. If you are vulnerable and weak, he will take advantage of you and devour you. He is a liar and deceiver and the father of lies. In the Gospel of John 8:44, it says, "When he speaks a lie, he speaks from his own resources, for he is a liar and the father of it."

I prayed to God and asked Him, was it because of unbelief, lack of faith, or lack of knowledge that my wife didn't get the healing? God revealed to me through His Word, books, TV ministries, and stories from well-respected apostles, bishops, pastors, etc. He said to me, "Look at all these great men of God. They have impacted millions of lives for the Kingdom of God. They had faith and they believed and did not receive the healing they asked and prayed for in their losses either." I went back and reviewed some of the stories of those great men of God; some had lost their wives, too. God let me know that it wasn't because of my unbelief, lack of faith or lack of knowledge that my wife didn't receive the healing.

The guilt I spoke about earlier in this chapter, God confirmed that too. He said, "Let it go; that is of the enemy." As I sit here sharing how great God is, the Holy Spirit has just given me another revelation. Abraham was the father of many nations. In Genesis 23, Abraham lost his wife Sarah in the land of Canaan, and Abraham came to mourn for Sarah and to weep for her. The Bible says in Romans 4:19, "Abraham was not weak in faith," and in Romans 4:20, "He did not waver at the promise of God through unbelief, but was strengthened in faith, giving glory to God, and being fully convinced what God had promised God was able to perform." How powerful is that! I will repeat that again. "Abraham was not weak in faith and did not waver through the promise of God through unbelief." Thank You, Holy Spirit, for that one; you knew I needed to hear that.

Jesus tells us in the Gospel of John that He has overcome the world, that there will be trials and tribulations, and that He is our peace. "These things I have spoken to you, that in Me you may have peace. In the world you will have tribulation; but be of good cheer, I have overcome the world." John 16:33 (NKJV)

In tribulation, death is included for everyone; we are all going to experience it. When it comes, put all your trust in Jesus. He is your peace and comforter; He defeated death. Know that your loved one is not permanently lost. They are with the Lord in heaven, and you will see them again.

Somehow, deep down inside I believe my wife did get a healing in another way. We can never know God's many hidden mysteries until the day we all come before Him to be judged. My wife is no longer in her sick body but rejoicing in heaven with the Lord. God's thoughts and ways are higher than our thoughts and ways. I also believe Psalms 115:3 where it says, "Our God is in heaven; He does whatever He pleases."

He is the King of Kings and the Lord of Lords. I will forever put my trust in the Lord. Every day that I wake up, I am one day closer to seeing my wife again. Thank You, Lord, for the blessing of eternal life with You forever. "Precious in the sight of the LORD is the death of His faithful servants." Psalms 116:15 (NIV)

PEOPLE

Now it was time to deal with people. Reality had just set in, and I had no idea what I was going to say to family members, friends and just people in general. Pearl was loved by so many people, and I knew it would be devastating news to her family and friends.

So, I decided not to tell co-workers on my job, only my immediate supervisors. I also requested that they tell none of my co-workers because I needed to have peace when I returned to work, whenever that was. I know that people can be very cruel and insensitive and say the wrong thing, thinking they are helping the situation. We live in a time where people in general are cruel, insensitive, mean spirited and quick to judge others. Knowing my co-workers there was a very good chance of this happening. I also didn't want my co-workers feeling sorry for me because that meant reliving my wife's passing and that would be like pouring salt on an open wound. Especially when I'm trying to heal from the most devastating loss ever in my life. I didn't want to relive it every day. My supervisors understood and respected my wishes. They gave me all the support I asked for and were there for anything I needed; they had my back. I did have one co-worker that I shared with, because I knew that I could trust that person. If I needed anything for the job, I could go to that person.

Friends were a lot tougher to deal with than the job, you can imagine. Thankfully, God gave me the words to say. My wife's spiritual and close friend in Detroit, who my wife always confided in for spiritual advice, was a blessing to me. She was a big help checking in on the family almost every day before and after my wife went home to be with the Lord. That helped me deal with my wife's besties and friends here in California who didn't take the news well. I found myself consoling them instead of the other way around.

Questions of how it happened and so forth arose. Like I said it was tough and took everything I had to comfort them without going off on some of them, but I managed to get through it. I guess I was being a hypocrite for feeling that way. Thinking she was my wife, and I should be the one who should be consoled, not really thinking that they loved her too. I was also feeling that they should have been more compassionate towards my feelings like my wife's spiritual friend in Detroit. The Holy Spirit gave me a revelation from Matthew 7:5: "Hypocrite! First remove the plank from your own eye, and then you will see clearly to remove the speck from your brother's eye." That cleared things up for me and those negative feelings and emotions I wanted to unleash on some of my wife's besties and friends in California went away. It was that still small voice from the Holy Spirit convicting me and correcting my behavior.

As I mentioned earlier in this chapter about friends, I thought I knew most of my wife's friends, but apparently not. I had people showing up at my doorstep besides neighbors, offering their condolences. I even had a person show up at my house unannounced that I never saw in my life. The person offered their condolences and told me that Pearl always prayed for her and with her. She said that those prayers got her through a lot of tough times. She just wanted me to know what a kind and loving person Pearlie was. I also took calls from my wife's phone because I already knew that she would be getting calls from people and more friends that I hadn't contacted because I didn't know their phone numbers. I thought about just having the service turned off on my wife's phone and that way I wouldn't have to deal with more people because I was overwhelmed. But I was ready this time. That still small voice was with me. I only struggled with one of those calls and that person must have sensed something because when that person's voice got louder mine got softer and I was more like a listener. That call ended on a good note. I was battle ready this time as I drew from *The Battle-Ready Prayer* by Aaron Hopson. It is a very powerful prayer. If you haven't read it do so; it's a must read and available in book form.

"He who has knowledge restrains and is careful with his words, and a man of understanding and wisdom has a cool spirit (self-control, and even temper)." Proverbs 17:27 (AMP)

Family was the toughest and I didn't even know where to begin. It was like I had a list in my mind who I was going to call first and one for my side of the family too. It didn't work out that way at all because the calls started. One phone call after another, I'd be on my cell phone, and someone would call on the house phone or on my wife's phone and leave a message. I felt like my phone was a 1-800 call center. It went on for a few days. I listened to family members grieve, get angry, question, and offer their condolences. It took a lot out of me, but I had my daughter with me. She appeared to be very strong or pretended to be. I think it hadn't really sunk in yet that our lives would never be the same again. We were so busy consoling other people, we didn't have time to think about grieving ourselves. I thank God that we got through the tough part, dealing with the people. I know if I didn't have God in my life when this unbelievable tragedy came upon my family, it would have been ugly and even more devastating.

I thank God for being with me every step of the way. I know He will never leave me nor forsake me. I have come to know even in those times when it feels like He is not there, He is. He hears us. Lastly, I pray that God will continue to give me the wisdom, knowledge, and understanding to continue to love people and see through our differences and be at peace with one another. Psalms 133:1 (NIV) says, "How good and pleasant it is when God's people live together in unity!" and Romans 12:18 (TPT), "Do your best to live as everybody's friend." In Jesus' name, amen.

GOD'S COMFORT

Finally, the ton of calls, text messages and emails I was receiving from family, friends, and people in general had dwindled down to not so many a day. Reality started to set in. My wife and best friend of almost 38 years was in heaven and was no longer with me and the kids. I cried out to God to help me and my children get through this tremendous, heart-breaking loss. "God heals the broken hearted and binds up their wounds, healing their pain and comforting their sorrow," Psalms 147:3 (AMP). I was trusting in God and His Word, period, no matter what. He is the only one that can bring comfort and peace in such a loss. I know His Word will not return void, meaning His Word will accomplish what it set out to do (see Isaiah 55:11).

It didn't take long for the God of comfort and peace to start answering my prayers and needs. People started showing up in my life bringing me comfort, peace, and love, saying the words I needed to hear, and the wisdom to heal and move forward. My pastor prayed for me and my family and did my wife's eulogy. God provided comfort and healing through one of my Bible study brothers in Christ who had lost his wife a few years ago. His testimony was powerful, and it helped me to understand some of the hurt and pain I was experiencing after a loss of this magnitude.

Even on my job God brought comfort and healing to me through my customers. I was at a mailbox one day talking to a customer and I didn't think anything of it at the time. But after I got through talking to that customer and moved on to the next house, God, out of the blue, gave me a revelation. He said, "Remember him, that customer you just spoke to? About five or six years ago he lost his wife, and you were consoling him at that same mailbox. And remember, your other customer a few doors down from him lost his wife too." I was thinking to myself like *Wow!! This is definitely the Holy Spirit speaking to me, right?* Then He said to me, "Look at both of them now. Just like I got them through their loss and brought them peace and comfort, I will do the same for you."

I had to think about it. The customer I had just spoken to was at peace now. The one child he had was now in high school; they were happy. And the other, God brought a new wife into his life. They both seemed to have new lives and be happy. I thought to myself again, *They had it worse than me. They had little ones to raise; all my kids are adults.* I said, *Thank you, Lord, for that blessing. I didn't have to raise three kids by myself.* I saw that customer again a few days later at the same mailbox. When I saw him, I walked up to him and said, "I want to thank you" and I think I hugged him. He said, "For what?" Then I told him about my wife's passing and going down memory lane with him when his wife passed. He offered his condolences and told me that he was there for me if I ever needed to talk. We were already friends but have become even closer friends after that day.

God continued bringing me comfort in other ways, through helping others in their time of need, of suffering, and pain. The past year I found myself comforting and ministering to others who were experiencing sickness, disease, and loss of loved ones too. It seemed like death, sickness and disease were all around me once I took the focus off myself. I realized that this is part of life and I'm not the only one that is going through pain and suffering. I found myself ministering to people I didn't even know. A young man was in the store parking lot as I was walking in, crying, sitting in his car with the door open. I asked him if he was okay, and could I get him anything. He said no. I asked him what was wrong. He said his sister was just in a car accident and might not make it. I asked him if he wanted to pray for his sister. He said yes. I asked him her name and we prayed for her. A few days later I walked past a lady and noticed she was in distress. I asked her if she was all right. She said her friend was just diagnosed with stage three breast cancer. I asked her did she want to pray for her friend. She said yes, and we prayed for her friend. They were both very grateful for the blessing of comfort and prayer.

God also brought comfort to me through TV ministry. I watched TBN every day. TBN ministered to me. I listened to people's testimonies from all around the world about how God has helped them through their trials and tribulations. I listened to messages from pastors and teachers that have helped me on my present journey. I also read a lot of books that have helped comfort me and answer some of the questions that I struggle with. There were three books that I read that really brought comfort to me and I could really relate to them. The first book I read was *Hope for the Hurting* by Tony Evans, and then *Divine Disruption* by the Tony Evans Family. The third book was *Crushing: God Turns Pressure into Power* by T.D. Jakes. These are powerful reads, and they contain wisdom that will bring you comfort in suffering.

I thank God for His comfort in suffering. He is the Father of mercies and God of all comfort. In Jesus' Mighty Name I pray, amen.

"All praise to God, the Father of our Lord Jesus Christ. God is our merciful Father and source of all comfort. He comforts us in all our troubles so that we can comfort others. When they are troubled, we will be able to give them the same comfort God has given us." 2 Corinthians 1:3-4 (NLT)

THOUGHTS

God was putting people in my life to help me move forward. But Satan was making a bid for my thoughts at the same time. I was having thoughts about Pearl. I started thinking of stuff like, was I a good husband to her? I started thinking about things I could have done to be a better husband; she deserved better. I thought about disagreements we had in our 38 years together. Was it my fault that she's no longer here? My mind was a complete mess at times. Satan had me feeling guilty and was telling me stuff like "If you would have been a better husband she would be still here today."

As I mentioned in one of the earlier chapters, Satan had me doubting my faith and belief in God after we didn't get the healing we prayed for my wife. It was tough, standing on the Word of God and not wavering and still losing her. I knew I had to get out of the doubt and unbelief thought pattern and feed my spirit with the Word of God. I was weak and devastated from losing my wife and very vulnerable and I knew it. I even cried at times. So I started meditating on the Word. Every time I would have one of those negative thoughts from Satan, I'd tell myself, "What does the Word of God say?"

I spent a lot of time in 2 Corinthians 10:4-5, meditating on spiritual warfare, the pulling down of strongholds, bringing all thoughts into captivity to obedience to Christ (See Ephesians 6:10-18). Putting on the full armor of God. Truth, righteousness, the gospel of peace, faith, salvation, and the Word of God. And Philippians 4:8-9, meditating (think) on whatever things are true, noble, just, pure, lovely, and of good report. I got a double dose of these scriptures on Thursday nights Men's Bible Study; that was a blessing too.

Even though I was feeding my spirit with the Word of God it was still a battle, but I didn't give up. I kept meditating on the Word, finding outlets like reading books, revisiting sermons from pastors like the late Apostle Frederick K.C. Price. He had an awesome series about the "Battle of the Mind, Our Thought Life and our Armor." That series was amazing, and it helped me tremendously. I go to that series all the time and read over my notes, and it just been a blessing to me.

Spiritual warfare is no joke. The devil comes for your thoughts the first thing in the morning. He knows if he can get inside your head early in the morning, he pretty much is in control of you for the rest of the day. I do my best each morning to feed on the Word of God. I want Him first in my life. He is the reason I'm alive. When I spend time with God in the morning, I have peace and I'm happy. When I start my day doing other things, I find my days a little more challenging. Little things may irritate me that would not have, had I started my day with the Word of God. Satan comes for you in the morning to create doubt and unbelief. He is rolling out the red carpet of temptation.

You can avoid that red carpet of temptation just like Jesus did when He was tempted by Satan in the wilderness. Jesus put on the whole armor of God. When Satan tempted Jesus in the Gospel of Matthew, Jesus responded with His armor, which is Bible knowledge. In Matthew 4:6, Jesus said to him, "It is written again, 'You shall not tempt the LORD your God.'" Just like Jesus, we need to put on all our armor, not some, but all, to combat Satan's fiery darts, the evil day, and the wiles of the devil. Satan has the same routine, and he keeps using it (see 2 Corinthians 2:11). That is targeting our mind to create doubt, fear, and unbelief so he can devour you.

He cannot devour you unless you let him. Read 1 Peter 5:8. It says the devil walks around like a roaring lion seeking whom he may devour. Notice it says like a roaring lion; my understanding is, pretending to be or acting like. Jesus is the Lion of Judah with all power and authority.

I know now more than ever that I need to stay focused and continue to feed my spirit with Bible knowledge, which is the Word of God, because the devil will be knocking at my door again. Just like in Luke 4:13, after the devil had finished tempting Jesus, he left Him until an opportune time. Meaning he was coming back. The devil will be back at my door too, looking to find a chink in my armor. He is looking for a way to get in to capture my thoughts. Jesus is our example We must stay battle ready, just like our Lord and Savior. In Jesus' name, amen.

THE FIRST TIME

As time started to pass, I began to sleep a little better. I thought that I was close to being normal again, but I wasn't anywhere near being normal again. Our anniversary was a few weeks away. It would be the first one without my partner. Naturally, I visited her memorial site for the first time and laid down some purple carnations, her favorite color.

Later that day my baby brother in Michigan called me. We talked for about 45 minutes, about the Dallas Cowboys and a bunch of other stuff. He asked me if I was okay because he knew where I went earlier, and it was my anniversary. My little brother rarely gives me good advice but that day he did. He said, "Big Bro, you know everything is going to be the first time for about a year." I was surprised and shocked that came out of his mouth. I was very grateful that he said it. I thanked my little brother; he made good sense, and I thanked God for using my little brother to minister to me.

God had just used my little brother to open my eyes and give me a revelation to use "the first time" as a healing process instead of a grieving process. Mother's Day I went to the memorial site with my middle son. After we left, it seemed like I had gotten a little stronger and I realized I was beginning to heal.

I prayed with my kids, ministered to them, and reminded them that Mom was in heaven, and this was just a memorial site for us to honor her memory. 1 Thessalonians 4:13 (NIV) says as believers we do not grieve as non-believers and Ecclesiastes 12:7 (AMP), says the dust that God made man's body from will return to the earth as it was, and the spirit will return to God who gave it. I also reminded them that we are spirits inside of a body and that we never die; only the body dies. I'm very glad and thankful that my kids are adults and can understand what

I prayed with them and ministered to them. If they were little kids, it would be a lot tougher for them to understand and I would have to teach them in stages as they got older so they could understand.

All my kids came home for Thanksgiving dinner, including my new grandson. This would be his first Thanksgiving ever and a blessing to our family. I cooked Thanksgiving dinner for my family. We had turkey and dressing, greens and cabbage, cornbread, salmon, short ribs, hot links, green beans. My daughter brought the mashed potatoes and gravy, and my son brought this special rice that everyone likes. We had pecan pie, lemon meringue pie, and a Patti Labelle sweet potato pie. We blessed the food and gave thanks to the Lord for allowing us to be together and thanked Him for our newest addition to the family (Jaylen Kai Daniels, my grandson), and we thanked the Lord for the time He gave us with my wife. This was our first Thanksgiving dinner without her, but we all knew that she was with us in spirit because of all the food and the love and togetherness in our house. She had always cooked Thanksgiving dinner and kept us close and together as a family. We ate, watched the football game, played PS4 games and enjoyed each other's company. I thank God for family and the blessing of healing broken hearts. In Jesus' name, amen.

"Now we do not want you to be uninformed, believers, about those who are asleep [in death], so that you will not grieve [for them] as the others do who have no hope [beyond this present life]. For if we believe that Jesus died and rose again [as in fact He did], even so God [in the same way-by raising them from the dead] will bring with Him those [believers] who have fallen asleep in Jesus. For we say this to you by the Lord's [own] word, that we who are still alive and remain until the coming of the Lord, will in no way precede [into His presence] those [believers who have fallen asleep [in death]. For the Lord Himself will come down from heaven with a shout of command, with the voice of the archangel and with the [blast of the] trumpet of God, and the dead in Christ will rise first. Then we who are alive and remain [on the earth] will simultaneously be caught up [raptured] together with them [the resurrected ones] in the clouds to meet the Lord in the air, and so we will always be with the Lord! Therefore, comfort and encourage one another with these words [concerning our reunion with believers who have died]." 1 Thessalonians 4:13-18 (AMP)

ANGRY WITH GOD

As I mentioned earlier in Chapter One, the day my wife went home to be with the Lord I cried out to God saying, "I'm still trusting You and I'm not angry with you." I also mentioned weeks later I would get angry at God as I grieved for my wife. I remember weeks later after all the dust had settled. The phone calls, text messages, and emails stopped, and everyone went back to their regular lives. I was no longer overwhelmed by people in general. Reality had really set in, and my life had changed completely. My partner of almost 38 years was gone and that made me angry.

I was so angry one day, not thinking before I opened my mouth, I let the devil deceive me through anger. I should have waited until I calmed down and not let my emotions and feelings get the best of me. Before I knew it, I cried out to God in anger, saying stuff like, "God, why didn't You take me instead of her?" over and over. I also said, "My kids needed their mother more than me." After that outburst the devil had my thoughts. I started thinking to myself, *I wish I could have taken her place so she could be here with the kids.* My mind was out of control; the devil had made me think that God was the one who took my wife. While I knew better, some man of God I was at that moment.

After I had calmed down and regained my spiritual composure, I repented to God and asked Him for His forgiveness. A few minutes later the Holy Spirit spoke to me, giving me a revelation, and He led me to John 10:10. "The thief does not come except to steal, and to kill, and to destroy. I have come that they may have life, and that they may have it more abundantly."

I knew the Word, but the devil found a chink in my armor. I was operating off my feelings and emotions. You can't fight a spiritual battle with the devil with feelings and emotions; you lose every time. You need all your armor, above all the shield of faith so you quench all the fiery darts of the devil (See Ephesians 6:16).

After getting back on track, I continued feeding on the word of God. I finally realized that I was angry with God because we did not get the healing for my wife, knowing that God had the power to heal her. I didn't ask God where He was when my wife passed. He was probably in the same place He was when His Son Jesus died. I think His Son has already answered that question. Jesus said that in the world we will have tribulation (John 16:33). I believe death, pain, and suffering are part of tribulation and everyday life.

I still have my moments of anger, but I'm no longer angry at God. I know it's Satan tempting me and creating doubt, which confuses us so he can capture your mind. The devil uses people we love, and things that have meaning to us, to deceive us. God doesn't tempt, the devil does (see James 1:13). I thank God for His patience and love, and for showing up in my hour of pain and suffering. Thank You, Lord, I'm trusting You to the end.

GRIEVING A PARTNER

"Therefore, a man shall leave his father and mother and be joined to his wife, and they shall become one flesh." Genesis 2:24 (NKJV)

Every time I did something that involved my wife, I grieved. I remember going to Costco to check our membership and return a bunch of blankets. I returned the blankets for a refund and Costco checked the membership renewal date for me. My wife was the one who took care of making sure that our membership stayed current. Costco verified our information and when they asked about card users, I let them know that my wife was no longer with us. They canceled all my wife's information. I was in pain; I wanted to start crying and I almost did. The grief I was feeling at that moment was almost unbearable.

The only reason I didn't cry was because I was in public.

I learned not only were we joined together as one flesh, we were joined together financially, physically, emotionally, spiritually, and intellectually. I had to remove my wife's name from so much that we shared. I made my children beneficiaries just in case something happened to me.

Every day I woke up I grieved. My wife loved our house; she left her touch all over it. Everywhere I went in the house there was something to remind me of her. So, I started rearranging things, putting certain things out of sight. That helped a little bit. My wife's clothes... I would pick up one of her scarfs to smell her scent on them just to help with the grief and the pain.

I started asking other friends and people that I knew who had lost their spouses. Some said they took all their belongings and donated them. Some said they left everything the way their spouse left it. I found out that everyone grieves differently. I know for me, certain things of my wife's bothered me, and some things didn't. Some of her things that bothered me I got rid of.

My wife's cell phone--it took me 11 months to shut it off because I already knew I would instantly grieve. I did not want to feel that same grief or pain I felt when her information was canceled off our Costco membership. I had to heal a little more and work myself up to it before I had the cell phone service turned off. I felt guilty, but I knew that thought came from Satan and I cast down that stronghold right away with the Word of God. I paid for a cell phone for 11 months that was just sitting there. It was worth it, not going through that grief and pain again, until I was able to deal with it.

I drive her car sometimes just to keep it up. Some days I'm okay with it and some days I have my moments of grief because I miss her and feel like she should be still here and driving this car. But I fight through it by standing on the Word of God. Eventually, as I get stronger, I will decide what to do with the vehicle. When you are grieving for a spouse or a loved one it's tough.

My wife had this little square porcelain business card that says, “Some people come into our lives, leave footprints on our hearts and we are never the same.” That one is a keeper for me, and I will pass that one down to my grandkids when they get old enough to understand what it means. I will share those stories about their grandmother and how she left footprints on our hearts. The Lord definitely blessed me with a wonderful wife for almost thirty-eight years. I thank Him for the blessing, and I thank Him for walking with me as I grieve. In Jesus’ Mighty Name I pray, amen. “A man who finds a wife finds a good thing and obtains favor from the LORD.” Proverbs 18:22 (CSB)

LONELINESS

"And the LORD God said, It is not good that man should be alone; I will make him a helper comparable to him." Genesis 2:18 (NKJV)

I battled loneliness. I spent a lot of time by myself on the weekdays after work. My kids were all adults; they had their own lives and their own places. My brother in-law lived with us, but it was still like I lived alone. When I would come home from work, he would be at work. When he got home, I was asleep. When I got up and went to work, he was asleep. We saw each other on Sunday after I came home from church. I found out you could live in the same house with someone and still feel lonely.

I spent time with my kids when I could. I spent the most time with my oldest son. He called and checked up on me the most. My kids would tell me to stay and not to go home to an empty house. I told them that I couldn't stay sometimes. I needed to go home because I must face it so I can heal. I had to get used to their mom not being there when I got home. It was hard but I had to do it. And I reminded them that their uncle was probably feeling the same way too. So, I had to go home because of him too and spend some time with him, because we only had time on Sundays to interact with each other. I didn't want him feeling like I was feeling.

The most difficult time I had experiencing loneliness was when I came home from work. That was hard, because every time I stuck the key in the door to unlock it, I thought about my wife. She would always yell, "Jake, is that you?" and I would say, "It's me" and walk in the house to the kitchen. She would always say, "Didn't you hear me calling you when you was sticking that key in the door?" I would say, "When you said 'Jake, is that you?' I said yeah." She would then say, "Oh, I didn't hear you, the next time speak up." She would smile and jokingly point to the knives on the cabinet and say, "You almost got cut because I didn't hear you." I'm like, "Okay, I'll yell a little louder next time." I would kiss her on the side of her face and get a mouthful of hair, because she always tilted her head to the right and downward. She thought that it was funny that I had hair in my mouth; she was a jokester. The days I didn't want hair in my mouth I just hugged her. She would then say, "You're squeezing me too tight" and I would say, "It's more of you to love."

We had an agreement: when I came home from work, she had a 30-minute rule. She gave me 30 minutes to unwind and take a shower, then come back downstairs to the kitchen. We sat across from each other at the kitchen island and ate dinner and I listened to her talk about the kids and whatever else she wanted to talk about or tell me. Sometimes I would stay upstairs a little longer and she would yell, "Where are you?" I would yell, "I'm doing something; I will be right down." So, the next day it would be like I was on punishment for not coming downstairs on time. When I walked through the door, she would say, "Sit down; eat your food before it gets cold." Then we'd eat and she would do all the talking and she would get her time back from the day before. When she finished talking, she would say, "Okay, you can go upstairs now and take your shower and unwind. I will meet you upstairs." That was our daily routine when I came home from work. That's what made it so hard for me to come home to an empty house.

I didn't want to put pressure on my kids to come home to the house they grew up in. I knew from experience that it was hard for them, just as much as it was for me. When they came over, they never stayed but for a minute. My youngest son would always tell me to come over and hang out because they had the extra room. My daughter is my fishing partner; she would always say on her way out, "Dad, let's plan a fishing trip." "Call me with the day that you want to go so I can take off work." My oldest, he would always have to get back because of traffic. I had a conversation with each one of them, I believe, on separate occasions. I wanted to let them know that I understand why they don't come to the house as much. I told them do not feel bad or worry about me, I'm okay and its okay to feel the way they do. I also let them know it's going to take them some time to heal and move on. As hard as it may be, I'm starting see them more now at the house as the months go by. I thank God for the blessing of helping my kids move forward in their young lives.

I prayed to God to take away the loneliness of being in the house by myself. Deuteronomy 31:6 (NIV) says "The LORD will never leave you nor forsake you." I prayed and meditated on that scripture and God heard me. It was amazing; a couple of days later I no longer felt lonely in my house when I was by myself. But I was still lonely for my wife; that feeling wasn't gone.

I fell asleep one night watching one of the gospel channels. I forgot which channel; I have cable and they go from 560 through 575. There are about five or six gospel channels in between 560 and 575. Well, anyway, to make a long story short, I woke up about 3:30 am and there was this elderly woman on TV. She was giving her testimony; she lost her husband over 40 years ago. She said she turned to God, and He took away her loneliness. She said every night she lay in bed for the past 40 years, she has never felt lonely. She feels like God has His arms wrapped around her and He is with her every minute of the day. It was a powerful testimony.

I still miss Pearl and am lonely for her companionship even though I know I will see her again in the next life. I'm getting better every day. My life is focused on following Jesus and doing work for the Kingdom and being obedient to God. Also, spending as much time with my family as possible until it's my turn to be absent from the body and present with the Lord. Loneliness is tough when you lose somebody that you've shared over half of your life with. It hasn't been easy, but if I'm walking, believing, trusting, and putting my faith in God, I know this too shall pass. I thank God for His mercy, forgiveness, love, patience, and guidance. He will never leave me nor forsake me. I'm trusting God, period.

OBEDIENCE TO GOD

I wrote this book out of pain and obedience to God. I was led by the Holy Spirit and given this assignment when I was suffering, in pain, angry, grieving, vulnerable, and going through the worst time of my life, which was losing my wife. I was feeling lower than low. When I say "out of pain and obedience to God" it was exactly that. I really didn't want to do it and I wasn't feeling it, because for a minute I was angry at God; I mentioned that earlier in the book. We didn't get the healing for my wife. I believed God; I put my all in Him. After losing Pearlie I was all mixed up.

I still trusted in God through it all because I knew as a follower and believer of Jesus Christ, He would be the only way that I could heal. As

I mentioned in one of the other chapters, I made up my mind to continue trusting God. The Holy Spirit brought to my attention that my testimony could help a lot of hurting people who were experiencing the same pain and suffering.

It seemed like every day as I went through my pain and suffering someone else was too. I was ministering to people who had lost loved ones just like I had. It was hard sometimes, but I endured. I was ministering to people who were diagnosed with internal stage 3 sicknesses. I was ministering to others and God was working on me. I finally got to the point where all the anger just went away. It was like I had a new spirit inside of me.

I believe that God was testing me, to see if I still believed and trusted in Him after my loss. In the Gospel of John, chapter fifteen, Jesus talks about obeying Him and staying true to His word and whatever we ask for it will be done for us. "If ye abide in me, and my words abide in you, ye shall ask what ye will, and it shall be done unto you." John 15:7 (KJV)

I thank God for walking with me through the

darkest time of my life. He was clearly with me. I couldn't have gotten through all this without Him. I completed the assignment He gave me, and I received healing through the writing of *Pearl.* It is remarkable how God used my pain and suffering and obedience to heal me. He also used my pain, suffering, and obedience to heal and comfort others in their pain and sufferings. Lastly, He used this earthly vessel to produce *Pearl* and send a message to the lost that are suffering and in pain. God is no respecter of persons; what He will do for one, He will do for another. God shows no favoritism or partiality. See Romans 2:11(NKJV).

I pray that whoever reads this book, God will deliver them from their pain and suffering. I give God all the praise, the glory, and the honor. In Jesus' name, amen.

"Yea, though I walk through the valley of the shadow of death, I will fear no evil; For You are with me; Your rod and Your staff, they comfort me." Psalms 23:4 (NKJV)

ACKNOWLEDGEMENTS

First, I want to thank God the Father, the Son, and the Holy Spirit for allowing me to be part of His plan. I thank Him for His grace and His mercy. I thank Him for holding on to me as I weathered the biggest storm of my life. I thank Him for loving me. I ask and pray that He continues to guide me and teach me in the way He wants me to go. In Jesus' name, amen.

I want to thank my children for being strong and for their love and support. You guys are truly amazing and I'm very proud of you. Always remember to keep God first in everything that you do. I love you guys more than you will ever know.

I want to thank the people that God sent to minister to me. I want to thank the people who shared their testimonies and stories with me. I want to thank the people that supported me with kind words and love. I want to thank those that supported me during my most vulnerable times of pain and suffering. Lastly, I can't forget my leaders at my job. You had my back; you were there every time I needed you. Thank you.

DEDICATION

I dedicate this book to Pearlie Bell Daniels, my wife and best friend for nearly 38 years. I thank her for her love and support and the life we shared together. I'm grateful for our three children and all the memories. I thank her for all the sacrifices she made for our family, keeping us together. She had a warm smile, a kind heart, and she loved people. I miss her dearly, but I know she's in heaven and I will see her again. Heaven is where I want to be when my time comes.

"Again, the kingdom of heaven is like a merchant seeking beautiful pearls, who, when he had found one pearl of great price, went, and sold all that he had and bought it." Matthew 13:45-46 (NKJV)

CPSIA information can be obtained
at www.ICGtesting.com
Printed in the USA
BVHW051921220223
659019BV00011B/111